Air, the Environment and Public Health

Anthony S. Kessel B.Sc., M.B.B.S., M.Phil., M.Sc., M.F.P.H., M.R.C.G.P.
Director of Public Health, Camden Primary Care Trust, London, and
Director, International Programme for Ethics, Public Health and Human Rights,
London School of Hygiene & Tropical Medicine

CAMBRIDGE
UNIVERSITY PRESS

The influence of public health was already in decline, though, after the introduction of the National Health Service (NHS) in 1948. This was partly because it was the period when Britain, like other developed countries, was going through the epidemiological transition which saw the burden of ill-health shift from communicable to non-communicable disease. This transition led to the view that public health surveillance would be of declining importance, as non-communicable disease was not seen as involving environmental issues. Also the Medical Officer of Health's management role was substantially reduced in 1948 because the NHS was organised nationally rather than at the level of the local authority, and the increasing professional autonomy of sanitary inspectors, and social workers in the post-war years finally led to the demise of the Medical Officer of Health in 1974.

The paradox has been that since this time there have been demands for a renaissance of public health, particularly under the banner of the 'new public health'. There is a variety of reasons why this arose in the closing years of the twentieth century, but two are of particular relevance to this book. The first of these is the renewed interest in the environment, and especially in air. Indeed, since the great smog episodes of the mid-twentieth-century, there has been an increasing awareness of the health risks from air pollution, and this, in turn, has underlain the development of clean air legislation and regulatory action. More recently there has been both a resurgence of communicable diseases, some of them airborne, on top of the rise of non-communicable diseases such as chronic respiratory disease and asthma, and new concerns with transport-related air pollution, acid rain and, now, changes in the global climate induced by greenhouse gas emissions.

The second reason is that the traditional biomedical model, with its reliance on scientific positivism and medical paternalism, has increasingly come under challenge. This has highlighted the latent tension between the two traditional public health roles, those of surveillance and protection of the population's health, and the management of the medical care system. When it was assumed that the public's, the state's and medicine's conceptions of health could be objectively defined and so were logically coincident, combining these two roles was rarely seen as problematic. But since the 1960s these constituencies have been seen as having potentially competing interests, reflecting their different perspectives on health; this plurality of vision now presents an ongoing and fundamental challenge to the single lens of biomedicine. Thus public health is open to a range of new interpretations, particularly concerning whether it should be under the control of organised medicine, and so is revisiting some of the debates and political struggles of the nineteenth century.

This book is set within this new era of controversy, and has both arisen from it and constitutes a critical reflection on it. In order to accomplish this, though, the book not only engages with contemporary concerns of relevance to public health but also breaks with the traditional academic mould, and so engages with issues of wider interest.

History and philosophy of science is now a well-established discipline but is relatively under-developed in relation to medicine and public health. The reasons for this are twofold. First, the science of medicine and public health is seen as applied, as practical, and so is often considered as less suitable, as well as less important, as a subject for analysis than the 'pure' sciences. Second, there has been a strong tendency for all academic disciplines to become more and more specialised and reductionist. So, although there has been an increase in the range of subjects seen as relevant to medicine, this has mainly resulted in a proliferation of subdisciplines, e.g. social history of medicine, medical ethics and medical sociology, and these have tended to be informed by the methods of the parent disciplines from which they derived.

Breaking out of this mould to develop interdisciplinary perspectives is difficult but essential if the new public health is to flourish, and Anthony Kessel has responded to the challenge. He has been able to do this because he brings to the task an unusual combination of qualifications, in medicine, public health, history and philosophy of science, and ethics. It has allowed him to interweave and combine a concern with the practical aspects of air and health, with theoretical issues, and, in doing so, to demonstrate the more general potential of this type of methodology. This breadth of vision has then enabled the exploration of a wide range of issues, including the changing relationship between humans and the natural environment (extending now to the unprecedented prospect of human-induced changes in global atmospheric composition and, thus, in world climatic–environmental conditions), the relevance of the humanities to public health education, research and practice, and the connections between global health and political philosophy. Hence, this book will appeal not only to practitioners of public health but to all those who are concerned with the future health of the planet and the development of interdisciplinary studies.

Dr David Greaves
Honorary Senior Lecturer in Medical Humanities
Centre for Philosophy Humanities and Law in Health Care
School of Health Science
University of Wales
Swansea
UK

Professor Anthony J. McMichael
Director, National Centre for Epidemiology and Population Health
The Australian National University
Canberra
Australia

Acknowledgements

This book has been several years in the writing, during which time a number of people have helped in different ways. Special thanks go to David Greaves and Tony McMichael, who have provided expert input and guidance throughout the period. I am indebted to Jeanelle de Gruchy for her observations on careful proofreading of the whole book, and to John Porter for the many hours of discussion that have informed my ideas. I am also grateful to the following for taking the time to comment on drafts at various stages: Andy Haines; Martin McKee; Virginia Berridge; Susannah Taylor; Dave Leon; Sari Kovats; Mike Ahern; Chris Watts; and Don Hill.

Many libraries were used for researching this book but I would like to express particular gratitude to the National Society for Clean Air and Environmental Protection for the generous manner in which I have been allowed to use their facilities.

Finally, this book would not have been possible without the encouragement and wisdom of my wife Elizabeth.

Abbreviations

AOSIS	Association of Small Island States
APHEA	Short-term effects of air pollution on health project
ASEAN	Association of South-East Asian Nations
BHHA	Barking and Havering Health Authority
BMJ	British Medical Journal
BS	Black smoke
CIAP	Committee for the Investigation of Atmospheric Pollution
COMEAP	Committee on the Medical Effects of Air Pollution
COP	Conference of the Parties
CSAS	Coal Smoke Abatement Society
DPH	Diploma in Public Health
DSIR	Department for Scientific and Industrial Research
EHO	Environmental Health Officer
EU	European Union
FCCC	Framework Convention on Climate Change
FCM	Faculty of Community Medicine (UK)
FPH	Faculty of Public Health (UK)
FPHM	Faculty of Public Health Medicine (UK)
GBH	General Board of Health
GCI	Global Commons Institute
GDP	Gross domestic product
GEC	Global environmental change
GIS	Geographical information system(s)
GP	General practitioner
Gt	Giga-tonne
HES	Hospital Episode Statistics
IPCC	Intergovernmental Panel on Climate Change
LAQM	Local Air Quality Management
LBBD	London Borough of Barking and Dagenham

LBH	London Borough of Havering
LGB	Local Government Board
MFPH	Membership of the Faculty of Public Health (UK)
MFPH Part 1	Membership of the Faculty of Public Health (UK) Part 1 examination
MFPH Part 2	Membership of the Faculty of Public Health (UK) Part 2 examination
MFPHM	Membership of the Faculty of Public Health Medicine (UK)
MFPHM Part 1	Membership of the Faculty of Public Health Medicine (UK) Part 1 examination
MFPHM Part 2	Membership of the Faculty of Public Health Medicine (UK) Part 2 examination
MOH	Medical Officer of Health
MP	Member of Parliament
NAQS	UK National Air Quality Strategy
NEHAP	United Kingdom National Environmental Health Action Plan
NHS	National Health Service
NO_2	Nitrogen dioxide
NO_x	Oxides of nitrogen
NIMBY	Not In My Back Yard
NSAC	National Smoke Abatement Committee
NSAS	National Smoke Abatement Society
OECD	Organisation of Economic Co-operation and Development
PCT	Primary Care Trust
$PM_{2.5}$	Particulate matter less than 2.5 micrometers (μm) in diameter
PM_{10}	Particulate matter less than 10 micrometers (μm) in diameter
QRA	Quantitative Risk Assessment
RCP	Royal College of Physicians (UK)
RCT	Randomised Clinical Trial
RMS	Royal Meteorological Society
RR	Relative Risk
SAC	Smoke Abatement Committee
SO_2	Sulphur dioxide
StHA	Strategic Health Authority
TB	Tuberculosis
TSP	Total Suspended Particulates
UK	United Kingdom
UN	United Nations
UNFCCC	United Nations Framework Convention on Climate Change
US	United States
USA	United States of America
WHO	World Health Organization

Introduction

This book traces conceptions of air and health from ancient civilisations to the present day, and explores the conceptions alongside historical developments in public health. Through examination of these changing relationships the book identifies and critically examines contemporary problems – scientific, philosophical and ethical – in public health theory and practice.

The introduction first explains why the theme of air and health was chosen, and then expands on the aims of the book. Next there is a discussion of some of the academic strengths and weaknesses posed by adopting an interdisciplinary approach, and one that spans such a long time-scale. Then, to place the book in context, there are introductory sections on what is meant by public health, environmental health and environmentalism. Although the public health focus is the UK, and a synopsis of the current situation in England and Wales is provided, international dimensions are also considered. Finally, outlines to the chapters are presented both as a guide to the shape of the book and also as a point of reference.

Why 'air and health'?

The environment appears to be making a comeback. After centuries of widespread environmental damage, attention is finally being directed to the importance of conservation to, and preservation of, the earth's natural resources.[1] Concern for valued natural resources – air, wild forests and endangered species – has become more acute due to recent fears that damage might be long-term or even irreversible. But how deep is the recent resurgence in interest in the natural environment, and does it matter what underpins it?

There is one form of rational and uncomplicated response to this question. Humans have harmed the environment and eventually this impacts back on mankind. Examples abound of how the damaged environment inevitably affects those who degraded it as well as those who did not: polluted water supplies causing birth defects; climatic disasters linked to global warming; and loss of plants with medicinal potential from destroyed rain forests [1]. A rising awareness of the consequences of environmental damage to human health and well-being has driven us to focus again on the environment and how we treat it. At last, recognition of human-induced environmental damage is being taken seriously.

But there is also another interpretation of the problem, based on related premises but offering alternative explanations. The argument here is that the root of the problem lies in the way humans relate to the natural environment, and how this has changed over time. Many ancient civilisations, as well as some more contemporary worldviews, picture mankind's relationship to the natural environment holistically, as part of an integrated whole. The environment has inherent value within such cultures and perspectives, rather than instrumental value for human needs and aspirations, whether these be related to health or purely aesthetic. Respecting the natural environment is an integral part of any such philosophy, not a belated add-on [2].

This second interpretation is linked to the belief that technological fixes will be insufficient to address the environmental problems of today and those of tomorrow. While technology will undoubtedly play a necessary part in efforts to attenuate current environmental damage (recycling, greener fuels, natural energy sources and so forth) and ameliorate future damage, alone it will not suffice. Only dramatic changes to the way humans live their lives, alongside a different relationship between mankind and the natural world, will secure the safety of the planet and its inhabitants. The shallow environmentalism of the modern West must be replaced by a deeper environmental commitment, requiring wholesale changes in Western behaviour and politics [3].

While these two approaches to the same environmental crisis are well recognised, exploration of the broader connections between the approaches and developments in the history of science and medicine has been relatively thin. In particular, it should be possible to look at historical changes in, say, understanding of human health and well-being, and see whether these changes shed light on interpretations of the causes of – and thereby possible solutions to – the environmental ill health of today.

It is feasible to go down this investigative route using a number of different ideas or themes. After all, the roots of contemporary environmental problems may be reflected in different historical developments and processes. For example, the origins of environmental ills might be linked to changes in leisure and travel patterns, and the importance these are perceived as having for health and psychological development. Or it might be useful to examine the history of the pharmaceutical industry and the impact that development of drugs based on natural substances has had on respect for the environment. Yet, as fruitful as such investigations might be, it is hard not to hold that leisure patterns and the pharmaceutical industry are in some respects too far removed from the environmental crisis to yield substantial and practical connections.

Instead, this book needed a theme closer to the natural world but still a health-related theme that could be historically followed against changing human relationships with the natural environment. Still, any one of a number of different themes could have been used, for instance water and health. Choice of theme, however, was somewhat dependent on the goals of this book and so it made sense to choose a health-related theme close to the area of medicine that has tended to tackle how the environment affects humans, namely close to public health. And with this goal in mind, the choice of theme for this book became more obvious: air and health.

Classically, human health has been associated with atmospheric air, from the harmonic humour of ancient civilisations, through the Old Testament's 'breath of life' and on to modern concerns about air pollution. Further, public health developed in the mid-nineteenth century against a backdrop of fears about the effects on the workforce of filthy air from unsanitary living conditions and factory smoke-filled skies. And attached to those fears was the charged debate over whether infectious diseases, the scourge of expanding economies of the nineteenth century, were transmitted by contagious persons or conveyed to individuals through the air as miasma.

So this book takes the theme of air and health, tracks conceptual changes in this theme against developments in public health and, in so doing, intends to illuminate the environmental problems now experienced.

Aims of this book

More specifically, this book has two main aims. The first is to explore historically the theme of air and health, and the relationship of this theme to developments in public health, particularly in England and Wales but also in other countries. Following this, the second aim is to use this exploration as a vehicle to examine critically generic issues in contemporary public health theory and practice, and to look at what these might tell us about the origins of today's environmental problems.

These aims present substantial challenges. Historiography, broadly speaking, can be based around two approaches: examining a particular idea or area in great depth over a specified (usually short) period; or taking a theme over a longer period and looking at links, for instance between ideas and practices. The former tends to be the preferred approach of academic historians, as attention to detail unravels the historical intricacies and helps inform how political, economic and other social processes shape change.

This book, however, adopts primarily the latter, largely because of the wider goal of using the theme to inform evaluation of contemporary issues. Attention to the fine historical details of specific events or periods can be fascinating and enlightening but it can also foster criticism that the results are largely of academic interest rather than practical utility.

Of course, from a historical perspective tracing an idea or a theme over more than two thousand years is no easy task. Not only is the time-frame huge but it encompasses vastly differing epochs, cultures and civilisations. Some areas are inevitably covered in less detail than others, and trends in themes can be difficult to identify and defend. Also, efforts to compare periods on such a large scale are inevitably open to criticism of failing to understand ideas, beliefs and events in relation to the context in which they appear. This book bears these warranted pitfalls in mind, alongside related tendencies to interpret historical events in the light of what is known today: so-called Whig historiography [4].

But the historiographic difficulties are certainly not the only ones; there are problems of definition. The notion of air, for example, has had diverse meanings, from the expired breath of an individual to a spiritual ether, or, in more modern times, the space that connects us as human beings and communities. And, as is more than familiar to students of public health or medical sociology, the term 'health' is notoriously difficult to pin down, notwithstanding the countless attempts to do so.

Further difficulties of definition include the array of perceptions of what public health may be, what constitutes the environment and what should be the appropriate subject matter of environmental health. To some, public health represents any collective effort to improve the health or well-being of the public; to others it refers to nearly two centuries of doctor-led professional activity geared towards advocacy and promotion of the health of communities [5]. Similarly, to some, environmental health is about how the environment impacts on human health, say through pollution of water supplies. But to others environmental health is really about the health of the environment, and nothing to do with how that affects mankind.

These last difficulties are returned to later in the introduction, with a synopsis of developments in public and environmental health, but a final intellectual problem needs mentioning. This is way this book mixes academic disciplines in tackling its subject.

History, science and philosophy: a critical blend

In addition to a historical perspective, this book draws on other disciplinary approaches – epidemiology, philosophy and ethics. Mixing in this way is sometimes considered academically challenging, to say the least. Some would argue that historians should grapple with original texts, scientists should do experiments and philosophers should stick to philosophising. Intellectual territory can be defended on the grounds that a historian would not be expected to conduct a clinical trial, almost inevitably can fall into a crisis of confidence and identity. 'If we're not doing *that*, what can we do?'

So we come to the latest restructuring of the NHS, as described in the following section. But returning to the introductory basis of this section, when considering developments in public health this book does indeed concentrate on England and Wales. However, most of the book covers theoretical aspects around changing conceptions of air and health that are of widespread relevance, and there is also further attention to international developments in public health in the conclusions.

There is no intention here to diminish the status of the histories of public health in many other countries, nor any desire to further promote the significance of the British experience, but it is largely to do with scope and application. It is just not possible to incorporate in any meaningful fashion the diverse histories of different countries, and this book also contains practical recommendations for UK public health based on its findings.

Recent changes to public health in the UK

Change is not new to public health in the UK. There have been numerous reorganisations since the profession's emergence in the mid-nineteenth century and now, 150 years later, public health is going through its latest, dramatic shift. As has often been the case before, changes to the public health function are part of a wider restructuring of the NHS.

The main driver of the recent changes has been the will to give more power to those working in primary care, power with regard to providing their own services, and also to commissioning hospital-based services (secondary, tertiary and quaternary services). In the pre-2001 system, health authorities[3] purchased virtually all services on behalf of the populations they represented, with public health departments located within those health authorities having a strong role in assessing the healthcare needs of their local populations. In an effort to contain spiralling costs the British Prime Minister Margaret Thatcher made failed attempts in the 1980s and 1990s to break the NHS monopoly and create a free, or freer, healthcare market.[4] The peculiar economic circumstances of the NHS made that difficult but the Labour Government of Tony Blair laid out a new direction at the turn of the millennium [26].

The new NHS would operate as a quasi-market with health authorities no longer the purchasers of services; instead, the buying would be carried out largely by new organisations called Primary Care Trusts (PCTs). Serving populations of around 100 000–200 000 people, by 2002/2003 PCTs would be responsible for purchasing approximately 75% of healthcare services for their local populations. As well as hoping to have an impact on costs, the financial power was being handed to the PCTs because it was thought that these organisations would know, and therefore serve, the needs of their local populations better than their predecessors. Rather than being predominantly administrative organisations, PCTs would be community-orientated and contain professional executive committees comprising key members of local primary care staff – general practitioners (GPs), district nurses, community midwives and so on. The PCTs would be more suitably placed to know what local patients need than the somewhat isolated former health authorities.

In 2001 the UK Government set out the details of the transition in *Shifting the Balance of Power* [27]. Some aspects have since progressed further, for instance the gradual transfer of commissioning activities to general practices from 2005. The important 2001 document, however, also laid out the implications for public health. While minimal public health teams would be based in new Strategic Health Authorities (StHAs),[5] most public health workers would be relocated to public health departments in PCTs. Because of the shortfall in public health skills these new departments would be smaller than their parent departments, and public health networks would be set up across StHA sectors to share skills and support thinly spread expertise.

These latest changes to the NHS, which are only now bedding down, offer the opportunity to reassess and reinvigorate public health in the UK. They also provide the chance for public health to 'go back to its roots'. Moving public health teams or departments to PCTs puts public health closer to communities, closer to assessment of their health needs and action to meet these needs. This responds to criticism that public health was too separate from the community. There is a very real threat, however, that public health resources, dissipated and fragmented, may be used up working on provision of health services and other politically directed agendas, rather than attending to social and environmental determinants of health. Demoralisation of a workforce tired of change, and with insecurity over its future, has led many to look elsewhere. With other related changes in the profession (examinations, non-medical status, and director posts; see Chapter 12, 'Conclusions and recommendations'), public health in the UK really is at a crossroads. During the course of this book contemporary and historical developments in public health are looked at in relation to changing conceptions of air and health.

Environmental health and environmentalism

In the decades leading up to the NHS restructuring described earlier, there has been a general heightened interest in the natural environment, manifest in public health through increasing epidemiological studies exploring links between the environment and human health: air pollution and climate change are obvious examples. But the environment, and environmental health, mean different things to different people.

Environmentalism refers to the broad ideology that gives the natural environment a more central place in the way we think and act. Although there has, since around the 1960s, been a growing leaning towards this kind of thinking in the Western world, the passion, commitment and intellectual component can vary substantially. While many share a desire that we care more for the natural environment by polluting and damaging less, far fewer hold the belief that the environment – and the way we treat it – are central to our philosophical, spiritual and political beliefs, and need to be prioritised accordingly [28]. In general terms, the former position demands relatively little of the individual and society (sometimes called 'shallow environmentalism') while the latter implores radical change ('deep environmentalism') [29]. Many individuals and the mainstream political parties might fit into the first category, but the second is more the domain of environmental philosophers, deep ecologists, pressure groups (such as Greenpeace and Friends of the Earth) and some of the political green parties [30].

Within public health these distinctions are similarly seen. What is referred to as environmental health is typically not about the health of the environment per se, but about environmental components of threats to human health. For example, a public health textbook of 1999 suggests that environmental health can be interpreted in two different ways. The first is the 'collective term for the administrative and regulatory functions carried out by the environmental health service in local authorities in the UK' and the second, drawing on the 1997 report of the UK Environmental Health Commission [31],[6] describes environmental health as:

> . . . the development of a wide understanding of the links and trade-offs between the environmental, social and economic factors that affect human health and the interventions needed to improve it.

Although this undoubtedly is a richer interpretation of environmental health, bringing in the broader links between socio-environmental conditions and health, it is still centred around human health. Today's environmental health problems are unsurprisingly described as including the maintenance and evolution of existing control systems, the re-emergence of key infectious diseases, and the lack of investment in infrastructure. The general perception of environmental health within public health circles is not fundamentally about the intrinsic health of the environment.

Herein is reflected one of the main themes that is elucidated and examined in this book. Has the relationship between mankind and the natural environment, in the way we see ourselves and understand our well-being and illness, changed to such a degree that our separation from the environment now only enables us to think of it as an instrument towards our own health?

Navigating the text

This book spans more than two millennia. The four parts of the book are purposefully divided, although they also broadly represent divisions of historical epochs.

Each part is created because it represents a reasonably demarcated conceptual relationship between air and health, or between air and the development of public health. References are provided for each chapter in turn, and a list of abbreviations is presented at the beginning.

Part I ('Whole Air') covers the period from early civilisations to the middle of the nineteenth century. Chapter 1 shows how in Greek medicine in particular – but also in other ancient medical systems as well as Roman medicine – air fitted into a holistic framework of health and illness that contained man and the environment, and was often imbued with elements of spirituality. Over a 2000-year period, the conception of air in medicine retained a sense of spirituality and holism, but also shifted markedly.

By the end of the Enlightenment and the birth of public health in Britain, air had become the medium around which a critical scientific debate about disease causation was taking place, and this debate was itself framed by a wider discussion about how the environment directed human social and moral development (see Chapter 2). While holding on to a sense of 'environmentalism', a narrower conception of air was beginning to appear, one which focused more specifically on air as part of the broader physical environment, and how air affected human health and disease.

In Part II ('Polluted Air') the changing conception of air and health is traced from the mid-nineteenth century until about 1970. As Western science began to underpin medical theory and practice during the second half of the nineteenth century, air became predominantly conceptualised as polluted air, and those involved in public health began to search for scientific evidence of links between bad air and ill health. Such evidence was not hard to demonstrate, but policy to improve the air failed to follow (see Chapter 3).

It is argued that this disconnected relationship between air pollution science and air pollution policy has continued through the twentieth century. The health effects of polluted air have long been recognised and repeatedly, although sporadically, illustrated. Legislation, however, to reduce air pollution levels and attenuate the adverse effects on health has generally either been slow to appear or failed to deliver, especially through the first half of the century (see Chapter 4).

The exceptional London smog of 1952 may have galvanised policy change, but changes in the structure and organisation of public health have distanced the public health worker from research, advocacy and 'ground-level' action. Air pollution levels may have fallen over the last decades of the twentieth century, but whether this has been driven by health concerns remains highly contestable (see Chapter 5).

Part III ('Air Pollution, Epidemiology and Public Health: Theoretical and Philosophical Considerations') follows chronologically and covers the decades up to the present, during which technological and methodological advances have seen air further reduced to components of polluted air around which the scientific search for associations with ill health has focused. To explore shifts in the conceptualisation of air and health a case study is used, involving a novel public health technique called Quantitative Risk Assessment (QRA). After a description of background developments in contemporary environmental health policy, the QRA that was undertaken is briefly described, including its scientific limitations (see Chapter 6). Part III, and indeed the book as a whole, focuses on atmospheric (outdoor) air pollution, and because of the limitations of scope is unable to cover areas such as tobacco and smoking. However, Chapter 6 does include a section that stresses the global importance of indoor air pollution and its relationship to poverty.

The following two chapters then consider what has shaped the emergence of QRA, factors external to the technique itself. These include developments in modern epidemiology, the focus on upstream risk factors and the lack of coherent epidemiological theory (see Chapter 7). Such issues in turn relate to philosophical and ethical questions about the nature of what constitutes evidence. Constraints on current epidemiological thinking and public health practice are discussed, as well as the implications for local and national policy (see Chapter 8).

The final and most recent conception of air and health is looked at in Part IV. In the form of global environmental change, and the approach to climate change in particular, air has become a space in which to rethink both the relationship between mankind and nature, and also the moral dimensions of public health theory and practice. Climate change presents an unparalleled example in which the health effects of Western lifestyles are borne by those at a distance in time and place. This raises fundamental questions about the geographical, temporal and moral boundaries of public (health) responsibilities, as well as the place of utilitarianism in public health theory.

An overview of the science of climate change and its effects on human health is initially presented. Next, the health dimensions of climate change are used as a basis to challenge utilitarianism as the moral foundation of public health (see Chapter 9). An alternative framework, social justice, is then put forward, and the emergence of 'climate justice' is explored; the related illustration of inequalities in health is considered. Yet social justice is not the only challenge to the domination of utilitarianism, and there has been criticism of the direction of modern Western moral philosophy. Other moral frameworks are examined through the example of virtue ethics and the perspectives of Ludwig Wittgenstein (see Chapter 10). The climate change debate, however, also highlights the lack of deep environmental thinking in Western and other contemporary cultures, and the significant challenges of environmental ethics are presented. If today's environmental problems have deep roots, then the solutions may require radical reform (see Chapter 11).

In the conclusions the findings of the book are summarised, and a critical look is taken at what can be inferred and drawn from these findings. Based on these interpretations some practical recommendations are made about the future of epidemiology and public health.

NOTES

1. Conservation involves active management, whereas preservation denotes keeping as wild.
2. This incorporated membership examinations, speciality training status and the establishment of the Faculty of Community Medicine (FCM). The FCM was later renamed the Faculty of Public Health Medicine (FPHM), within the Royal College of Physicians (RCP). In 2003 it was renamed the Faculty of Public Health (FPH).
3. There were about 100 health authorities in England and Wales, each serving a population of, approximately, between 250 000 and 450 000 people.
4. In a rather unusual economic situation the NHS is virtually both the sole purchaser (monopsony) and the sole provider (monopoly) of health care.
5. Bodies with responsibilities mainly around strategic planning and performance management of the NHS. A StHA might typically contain six to eight PCTs.
6. A body conceived by the UK Chartered Institute of Environmental Health.

REFERENCES

1. J. M. Last, Global change: ozone depletion, greenhouse warming, and public health. *Annu. Rev. Publ. Health,* **14** (1993), 115–32.
2. J. R. Des Jardins, *Environmental Ethics: An introduction to environmental philosophy* (Belmont, CA: Wadsworth, 1997).
3. J. Chesworth, ed., *The Ecology of Health: Identifying issues and alternatives* (Thousand Oaks, CA: Sage, 1996).
4. V. Berridge, Historical research. In N. Fulop, P. Allen, A. Clarke and N. Black, eds., *Studying the Organisation and Delivery of Health Services* (London: Routledge, 2001), pp. 140–53.
5. D. Porter, *Health, Civilization and the State: A history of public health from ancient to modern times* (London: Routledge, 1999).
6. F. Engels, *The Condition of the Working Class in England in 1844* (Moscow: Progress Publishers, 1973). [First published in German in 1844. First published in English in 1888.]
7. P. Bowler, *Charles Darwin: The man and his influence* (Oxford: Blackwell, 1990).
8. A. F. Chalmers, *What is this Thing Called Science?* (London: Open University Press, 1982).
9. R. Harre, *The Philosophies of Science: An introductory survey* (Oxford: Oxford University Press, 1972).
10. W. H. Newton-Smith, *The Rationality of Science* (London: Routledge & Kegan Paul, 1981).
11. H. Waitzkin, The social origins of illness: a neglected history. *Int. J. Health Serv.,* **11**: 1 (1981), 77–102.
12. J. Rawls, *A Theory of Justice* (Oxford: Oxford University Press, 1999).
13. A. Sen, *Inequality Re-examined* (Oxford: Oxford University Press, 1995).
14. A. MacIntyre, *Whose Justice? Which Rationality?* (Notre Dame, IN: University of Notre Dame Press, 1988).
15. A. J. McMichael, *Planetary Overload: Global environmental change and the health of the human species* (Cambridge: Cambridge University Press, 1995).

16. G. Rosen, *A History of Public Health*, expanded edn (Baltimore: Johns Hopkins University Press, 1993).

17. D. Porter, *Health, Civilization and the State: A history of public health from ancient to modern times* (London: Routledge, 1999), p. 4.

18. Quoted in T. H. Tulchinsky and E. A. Varavikova, *The New Public Health: An introduction for the 21st century* (San Diego: Academic Press, 2000), p. 70.

19. Committee for the Study of the Future of Public Health, *The Future of Public Health* (Washington, DC: National Academy Press, 1988).

20. Committee of Inquiry into the Future Development of the Public Health Function, *Public Health in England* (London: HMSO, 1988).

21. Faculty of Public Health Medicine (UK), *A career in public health.* www.fphm.org.uk/careers_in_public_health/careers.shtml (accessed 6 May 2005).

22. R. Beaglehole and R. Bonita, *Public Health at the Crossroads: Achievements and prospects* (Cambridge: Cambridge University Press, 1997).

23. C. Hamlin and S. Sheard, Revolutions in public health: 1848, and 1998? *BMJ*, **317** (1998), 587–91.

24. E. Fee and R. M. Acheson, eds., *A History of Education in Public Health: Health that mocks the doctors' rules* (Oxford: Oxford University Press, 1991).

25. J. Lewis, *What Price Community Medicine? The philosophy, practice and politics of public health since 1919* (Brighton: Wheatsheaf, 1986).

26. Department of Health, *The NHS Plan: A plan for investment; a plan for reform* (London: HMSO, 2000).

27. —, *Shifting the Balance of Power* (London: HMSO, 2001).

28. P. Singer, *Practical Ethics* (Cambridge: Cambridge University Press, 1999).

29. R. Elliot, Environmental ethics. In P. Singer, ed., *A Companion to Ethics* (Oxford: Blackwell, 1993), pp. 284–93.

30. N. Low, ed., *Global Ethics and Environment* (London: Routledge, 1999).

31. G. Jukes, Environmental health perspectives. In S. Griffiths and D. J. Hunter, eds., *Perspectives in Public Health* (Oxford: Radcliffe Medical Press, 1999), pp. 199–200.

Part I

Whole Air

Overview of Part I

Medical ideas and remedies began to be written down about 2000 BC, and from the records of these ancient civilisations onwards appear beliefs about health and illness, cause and cure. The aim of this part of the book is to explore the conceptions of air within different medical systems – from ancient civilisations to the bacteriological era of the end of the nineteenth century – and the meanings and implications of these conceptions.

Given that this book intends to use the theme of air and health to explore philosophical issues relevant to contemporary public health, the chapters in this part will necessarily focus on medical systems felt to be most pertinent, namely Greek medicine and scientific medicine in England around the time of Edwin Chadwick. It will then be possible later in the book to examine issues in modern Western medicine with reference to the medical systems of thought explicated in this chapter. It is not possible, given limitations of scope, to consider conceptions of air in all medical systems.

1

Early conceptions of air and health

Definitions and relationships: air, health and disease

There are no uniformly accepted definitions of health or disease, as these terms are culturally defined, and the meanings vary according to the context [1]. Even the present-day Western understanding of health is hard to pin down. While for some it may mean simply absence of disease, others consider the presence of health to be something more than the absence of a negative connotation. This is reflected in the idealistic and heavily used 1946 definition provided by the WHO [2], still often perceived as the founding block of many modern public health programmes:

Health is a state of complete physical, mental and social well-being, and not merely the absence of disease or infirmity.

As a word, 'disease' is no less problematic. Etymologically, the word stems from 'dis-ease', literally meaning not at ease, suggesting a departure from normal bodily functioning [3]. But this interpretation leads readily to a biomedical model of health which, though continuing to be the dominant model in modern Western societies, remains only one among many existing models [4].

In a similar vein, the term 'medicine' immediately throws up a picture of scientific medicine, its theory and practice: disease aetiology and pathophysiology based on the biomedical sciences, and treatment – scientifically tested and evaluated – provided by practitioners trained by (and monitored to standards set by) a profession. But a huge range of healing practices exist across cultures today, and historically, to which it is inappropriate to ascribe any collective term [5, 6].

Although there are no cross-cultural definitions of air, health or medicine, broad ideas have remained consistent between cultures distanced in time and place. These are beliefs about a general notion of well-being, departures from well-being, beliefs about the reasons for such departures (natural causation theories or personal/superhuman causation theories) and what should be done to better understand the departures and improve well-being.

Within these broad sets of beliefs air has consistently had a place. The meaning or understanding of air naturally varies with the context and includes: air as the external environment; air as wind; air temperature; air as climate; inspired air; expired air; components of air; and polluted air. However, within these different interpretations, certain conceptual themes emerge and are explored in this chapter:

• air as the natural or supernatural environment in relation to well-being/health;
• air as the natural or supernatural environment in relation to illness/disease;
• air as part of healing.

Air in the medical systems of ancient civilisations

The place of air in early societies can be seen first in the medical belief systems of ancient civilisations or cultures, three of which capture the conceptual space occupied by air: Egyptian, Chinese and Judeo-Christian.

In parallel with Mesopotamia, the Egyptian empire rose around 2000–3000 BC. The famous Smith papyrus, one of the oldest written records of medical thought available, speaks of the existence of surgery, an empirical component to medicine, and also a magico-religious element. Beneath the high priest Imhotep was a hierarchy of physicians, specialised in certain diseases or body parts, and health was associated with correct living and being at peace with the gods and the spirits of the dead. Illness was a matter of imbalance which could be restored to equilibrium by supplication, spells and rituals. Many remedies were concoctions of animal products, vegetables and minerals [7].

In the medical belief system of ancient Egypt, air had a formative place in two different ways. First, as part of an explanatory physiological system, air was thought to be contained in one of a mesh of vessels emerging from the heart, others carrying blood, urine, semen, tears and solid wastes. This conception – air as part of the physiological functioning of the body – was not exclusive to the Egyptians and is discussed further in the section on Greek medicine.

The second conception of air in Egyptian medical thought can be seen in the belief that 'life lay in breath'. This notion encapsulates a special, supernatural, place for air in the creation, or sustenance, of life. Such a concept was not unfamiliar to other early medical belief systems. For example, for ancient Chinese medicine and natural philosophy one of the two fundamental entities of nature is *qi*. Although impossible to translate, *qi* has been variously interpreted as air, vapours or life-force, that which for living beings is the vital energy sustaining life itself.[1] Ho and Lisowski suggest that *qi* can be thought of as the instrument 'composing all forms from below, and the tools and raw material with which all things are made', something not dissimilar to the Greek idea of *pneuma* (see below) or the modern concept of matter energy [8]. *Qi* permeates the wider cosmos, reflecting further

Chinese natural philosophical similarities with Greco-Roman thought, that the human body represents a microcosm of nature and society, and that demarcation makes no sense:

Unlike in Europe, science and the humanities have never parted company in traditional China, where every conceivable object or phenomenon, from astronomy to astrology, from alchemy to magic, from ethics to politics, and from philosophy to the art of healing, was considered to operate under the same principles of *li, qi* and *shu* [9].[2]

In early Judeo-Christian culture the concept of the breath of life, contained within a religious framework, is also evident. Christians accepted Greek doctrines of humours and temperaments but it was God who created healthy balance and who allowed unhealthy imbalance. Ideas of *pneuma* were similarly accepted but imbued with animistic vitalism,[3] with life produced by the Spirit of God breathed into inanimate clay. Alongside a strong emphasis on hygiene in Jewish medicine (physical cleanliness bespoke spiritual purity) [7], disease was essentially understood as an expression of the wrath of God, with evil connotations, and could therefore only be remedied by prayer, sacrifice and moral reform [10]. Judeo-Christian medicine held firmly to the spiritual dimension of health and disease that later Greek medicine had begun to dispel.

The contemporary author Jim Crace depicts this particularly well in his award-winning novel *Quarantine*. Although a piece of fiction, this carefully researched book captures both the spiritual connection between air and the creation or sustenance of life, and also the association of diseased air with evil. In one particular scene Jesus brings back to life the wicked trader Musa by removing the fevered evil air from his body [11]; unknown to Jesus at the time, Musa will become his nemesis later in the story:

He laid his hand on Musa's chest and pressed so that the devil's air expressed itself and filled the tent with the odour of his fever . . .

So, ancient civilisations indicate that the earliest conceptions of air in medicine relate to two intertwined notions. The first is the place of air in understanding the physiological functioning of the body, and the second embodies the spiritual significance of air as a life-giver. The overlapping nature of these conceptions reflects the inseparable understanding of mind, body and spirit, which is also seen in the best known of ancient medical systems, Greek medicine.

Greek medicine

As authors have pointed out, it is something of a myth to consider Greek medicine as a unified medical system, widely accepted at the time. Instead, there were different theories and practices competing in the Greek medical market place, and the individual citizen was free, cost permitting, to pick and choose between alternatives [12].

It is, however, reasonable to say that the first substantial medical texts were penned by the Greeks, and collectively known as the Hippocratic Corpus. These 62 books – and indeed those of rival natural philosophy-based medical theories of the time – represented, in the main, a departure from belief systems based on supernatural ideas of disease causation, such as expressions of the wrath of the gods [13].

Replacing this was the first rational medical theory, based on understanding humans as part of nature and illness as a natural phenomenon ('naturalism'), complemented by an ontological view of diseases as specific entities best understood through empirical observation [4]. But while the physician's role was expected to be that of observer and treater, it was the domain of the natural philosopher to explain the place of mankind in the universe and how the functioning of the human body (microcosmos) reflected the workings of the all-encompassing macrocosmos.

A result of this division saw physicians uncomfortably trying to marry their observational findings with existing natural philosophical theory. And, although Greek medicine had a firm rational basis, also evident were strong traces of supernaturalism, sometimes overlooked when Greek medicine is held up today as the early bedrock of Western scientific medicine [7].

Nevertheless, the extensive influence of Greek medicine for the following two millennia is unquestionable, making it important to explore closely the relationship between air and health in the Hellenistic period. Given the breadth of Greek medical writing it is impossible to look at all authors, and instead this chapter now focuses mainly on Hippocrates, Plato and Aristotle.

Air as the natural environment, and its effects on health

Written mostly between 430 and 330 BC, the Hippocratic Corpus is composed of 62 books covering various aspects of medical thought, including ideas about health, disease, prevention and cure. Originating out of the Hippocratic medical school on the island of Cos, the books were written in Ionian dialect by a number of different authors, and there is uncertainty over which, if any, were written by Hippocrates himself (Fig. 1.1) [13].

Despite this, Hippocrates was almost certainly a physician of high standing and it is likely that he oversaw much of the writing. His reputation was wide, the king of Persia famously asking his wisdom in an unusual case of love-sickness and, centuries later, the famous Roman physician Galen recounted Hippocrates' curing of an Athenian plague through bonfires lit to purify the air [14].

Figure 1.1   Engraving of a Classical Greek bust of Hippocrates (George Bernard/Science Photo Library).

For the less-noted Greek physician, local reputation as a successful healer was important, as income was dependent on demand. Although a few physicians were paid by the state and resided long-term in a city, most were itinerant, travelling from town to town in the hope of business. Within this peripatetic life lay a second related role of the physician: prediction of the health of a town based on its location and surroundings. This was useful not just in relation to anticipation of health problems that might affect the inhabitants, but also when physicians were called upon to assist in the siting of a new settlement [15].

In contrast to internal causes of disease, air – in the general sense of what comprises the atmosphere – was considered a possible direct external cause of disease. In *Breaths* Hippocrates [16] comments that 'it has been said that all living things participate to a large extent in air. After this, it must be remarked that it is likely that diseases come about from no other source than this . . .'

This theme is elaborated further in *The Nature of Man* [17]. Although contradicting the quotation above in suggesting that air alone may not be responsible for all diseases (reflecting that different treatises were written by different authors), the passage reiterates the importance of air:

Some diseases are produced by the manner of life that is followed; others by the life-giving air we breathe. That there are these two types may be demonstrated in the following way. When a large number of people all catch the same disease at the same time, the cause must be ascribed to something common to all and which they all use; in other words to what they all breathe.

In the same book the author continues shortly after by suggesting the means by which air may be responsible [18], and also a way towards improvement of health:

When an epidemic of one particular disease is established, it is evident that it is not the regimen but the air breathed which is responsible. Plainly, the air must be harmful because of some morbid secretion which it contains. Your advice to patients at such a time should be not to alter the regimen since this is not to blame . . . Care should be taken that the amount of air breathed should be as small as possible and as unfamiliar as possible.

However, probably the clearest indication of the importance placed on air as the external environment comes in the introductory section of *Airs, Waters, and Places*. Written in a style suggesting use perhaps both as a lecture and as a guidebook for physicians, this Hippocratic text [19] asserts the overall relevance of the seasons, air temperature and winds:

Whoever would study medicine aright must learn of the following subjects. First he must consider the effect of each of the seasons of the year and the differences between them. Secondly he must study the warm and cold winds, both those which are common to every country, and those peculiar to a particular locality. Lastly, the effect of water on the health must not be forgotten . . . When, therefore, a physician comes to a district previously unknown to him, he should consider both its situation and its aspect to the winds. The effect of any town upon the health of its population varies according as it faces north or south, east or west. This is of the greatest importance.

Following from the general comments implicating air as an external cause of disease, more specific relationships between aspects of air and disease causation can be traced. For example, in *Aphorisms* (one of the strangest Hippocratic texts, composed of hundreds of individual pointers about medicine) [20] Hippocrates asserts that the 'changes of the seasons are especially productive of disease, as are great fluctuations of heat or cold within the seasons.'

Also, although all diseases occur during all seasons, 'some of them [occur] more frequently . . . and are of greater severity at certain times' [21]. And to predict the seasons, the physician needs to be both in the 'business of the meteorologist' and must also [22] 'learn that astronomy plays a very important part in medicine since the changes in the seasons produce changes in diseases.'

Finally, air is perhaps most vividly portrayed as an external factor affecting disease when represented by wind. A Hippocratic author suggests [23] that when a district has cold prevailing winds from the quarter between north-west and north-east, and the water supply is hard and cold and usually brackish, the inhabitants will be 'sturdy and lean, tend to constipation, their bowels being intractable, but their chests will move easily.'

Consequent to these winds, inhabitants of districts 'that face east are likely to be healthier than those facing north or south even if such places are only a furlong apart' [24]. The most troublesome winds, however, are the north and south winds [25], cited as responsible for specific health problems:

South winds cause deafness, misty vision, headache, sluggishness and a relaxed condition of the body . . . The north wind brings coughs, sore throats, constipation, retention of urine, accompanied by rigors, pains in the sides and breast.

Air as the supernatural environment, and its effects on health

As outlined above, it was important for a physician to be able to use astronomical skills to assess seasonality, and to combine this with an understanding or a prediction of climate, in order to foretell health concerns. But the Hippocratic texts did not represent a complete departure from unnatural explanations of health and disease.

Although the gods were generally no longer implicated, the physician was advised to take into consideration a supernatural element akin to present-day astrology. Sometimes this seemed to be mainly as an indication of seasonality, for instance in *Epidemics* [26], when it is observed that at the time of Arcturus, if southerly rains continue until the equinox, then 'under such circumstances, cases of paralysis started to appear during the winter and became common, causing an epidemic.' But on other occasions [27] the link is less obvious:

Now let us consider the seasons and let us predict whether it is going to be a healthy or unhealthy year. It is most likely to be healthy if the signs observed at the rising and the setting of the stars occur normally, when there is rain in the autumn, when the winter is moderate being neither too mild nor excessively cold, and when rain falls seasonably in spring and late summer.

The uncertainty over the strength of association with the stars is evident as *Airs, Waters, Places*, from which the section above comes, later advises more specifically [28] that the 'most dangerous times are the two solstices, especially mid-summer, and the equinoxes' and care 'must also be taken at the rising of certain stars, particularly the Dog Star and Arcturus.'

This confused and incomplete transition from unnatural to natural explanations is perhaps most clearly apparent in the book devoted to the 'sacred disease', now known as epilepsy [29]. Here the author seems keen to invoke an alternative explanation for the disease most traditionally attributed to divine origin but cannot quite make the leap:

This so-called 'sacred disease' is due to the same causes as all other diseases, to the things we see come and go, the cold and the sun too, the changing and inconsistent winds. These things are divine so that there is no need to regard this disease as any more divine than any other; all are alike divine and all human. Each has its own nature and character and there is nothing in any disease which is unintelligible or insusceptible to treatment.

Air and the causation of specific diseases

Despite lack of consistency over the metaphysical basis of disease, causal factors were often cited as contributory to specific diseases, and epilepsy again provides a good example. Continuity of thought is evident if one looks firstly at what Plato said about epilepsy, and then looks back at what Hippocrates suggested two centuries earlier.

Plato's *Timaeus* is an unusual text, almost a stand-alone book, which has been lauded by some as providing important insight into Plato's cosmology as well as his views on health and disease [30, 31], but has also been criticised by others as confusing, inconsistent and poorly conceived [32].

The text itself takes the form of a dialogue between four characters, one of whom, Socrates, is expecting a reciprocal lecture to his own, *The Republic*, given the previous day. After a brief introductory section, the character of Timaeus provides Socrates with a lengthy monologue and, although it is mentioned that the other characters will speak later, they never do, suggesting an unfinished piece [33].

What we are left with is a fascinating, albeit difficult-to-follow, polemic on the nature of the universe (cosmos or macrocosmos, made of a World Soul and World Body), modelled by the divine artisan (Demiurge) from the cosmic paradigm (or Ideal) called The Living Animal. Given that the Demiurge is good, his product must be good in itself – ordered, intelligent, beautiful, and in perfect balance or harmony. It has been claimed that Plato's form of 'holism' lies at the roots of modern ecological thinking [34].

Naturally, in a text concerned with the cosmological order, the place of mankind is integral. Plato saw Earth and the natural world as part of the cosmos, with each human being portrayed as a microcosmos, constituted by the same principles that govern the cosmos. Harmony and balance are the natural state and allow health, while imbalance and disorder result in illness and disease. It is worth mentioning here the link with ethics, for Plato similarly felt that microcosmic imbalance caused emotional perturbation, resulting in loss of moral propriety, or the inability to make good judgements – the inability to lead a good life [35].

So, drawing on previous ideas about humours and elements, Plato invoked air – or imperfections in the nature or distribution of air – as causal of disease. For example, a third kind of serum (the first two being blood and another derived from black and acid bile) involves air and is a product of the dissolution from new and tender flesh, the decomposition of tender flesh intermixed with air described as 'white phlegm'. This may be responsible for skin problems [36]:

> White phlegm also is dangerous when it is blocked inside because of the air in its bubbles; but when it has air-vents outside the body it is milder, although it marks the body with spots by breeding white scabs and tethers and the maladies akin thereto.

More specifically, Plato goes on to suggest that when the flow of air through the lungs is blocked, white phlegm, in combination with bile, may cause 'countless diseases of a painful kind' and rotting of those body parts 'deprived of respiration' [37]. However, the most direct causal relationship is once again suggested in connection with the sacred disease. Following a necessary passage explaining the route of inspired air (first to the brain, then most to the stomach and some to the lungs and blood vessels) Plato asserts that air must be continually moving and 'if it remains still and is left behind in some part of the body, then that part becomes powerless' [38]. Then, most vividly, Plato links air, white phlegm and vessel blockage with the striking symptoms of an epileptic fit [39]:

> Therefore, when the blood-vessels are shut up from this supply of air by the accumulation of phlegm and thus cannot afford it passage, the patient loses his voice and wits. The hands become powerless and move convulsively for the blood can no longer maintain its customary flow.

Air as *pneuma*

The doctrine of pneumatism has a long ancestry and appears in various forms. According to Phillips, the medical sect of pneumatists, philosophically aligned with the Stoics, was founded around the middle of the first century BC, but the general notion of *pneuma* can be historically traced further back and was more non-specifically affiliated with the developing tradition of science [14].

In Coan medicine, for example, the concept of *pneuma*, or 'vital air', developed under the influence of natural philosophers such as Diogenes of Apollonia, and is well documented in the Hippocratic Corpus. *The Sacred Disease* represents air as responsible for consciousness or intelligence and, as discussed above, epileptic seizures were thought to result from the blockage of air (or *pneumato*) in vessels within the body.

In the *Timaeus* Plato similarly described *pneuma* as causing diseases such as tetanus, as well as pulmonary complaints and pleurisy, but his student Aristotle equated air, and breathing, more firmly with the spiritual notion of soul. In *On Breath* Aristotle theorises that breath, maintained and increased by nutriment, is the purest of all substances.

Aristotle wonders whether breath (divided into innate breath and inspired breath) is different to external air [40], but suggests that respiration has its 'motive principle from the inward parts' although he is unsure 'whether we ought to call this principle a power of the soul, the soul, or some other combination of bodies . . .' However, he clearly links air with spirit or soul [41], although he speculates over its relative contribution:

But if the soul resides in this air, the air is at any rate a neutral substance. Surely, if it becomes animate, or becomes soul, it suffers some change or alteration . . . air is not the whole of soul but is something which contributes to this potentiality . . .

Some time later the pneumatic school incorporated and developed these ideas into a concept of *pneuma*, or spirit, which in both the universe and man bound everything together, any alteration causing illness. *Pneuma* was seen as a fifth element which flowed through the arteries, sustaining vitality [42]. In the second century of the Common Era, the famous Roman physician Galen, who held Hippocrates in the highest regard, drew on the idea of *pneuma* as the life breath of the cosmos, while also believing that inside the body atmospheric air mixed with blood to form the vital spirits (or *pneuma*) that were responsible for creating the pulsative power within arteries. Galen is looked at below.

Air and the balance required for health

As mentioned in previous sections, later Greek medicine held balance within the body as key to health; disturbance of this balance resulting in disease. The causes of perturbations of equilibrium were explained in rational and natural terms, moving away from previous beliefs about the supernatural basis of illness.

There were, however, a number of different theories regarding what exactly was supposed to be in balance – elements, humours, a combination of these or something else. But within all these categories air had an integral place, either as an element in itself, as part of a humour, or as the external natural environment.

As Lund describes, the older philosophers of the Ionic school had regarded one single element as forming the substance of things, but the Pythagorean and Sicilian schools of medicine based a system on all four elements (earth, air, fire and water), and from this arose the doctrine of mixture or 'crasis' in the body [43]. The variety of ideas based on such a system is again well represented by Hippocrates and Plato.

Hippocratic thought is perhaps the most complex. In *Nature of Man*, the author refutes monism and instead introduces an elaborate theory correlating the four elements (earth, air, fire, water) with the four basic humours (black bile, blood, yellow bile, phlegm) and the four temperaments (melancholy, sanguine, choleric, phlegmatic), and further linking these with the four seasons, the four stages of man (infancy, youth, adulthood, old age), as well as the four primary qualities of hot, cold, dry and wet. Diseases were explained as a consequence of specific alterations of ratios within this finely tuned matrix.[4] Health, on the other hand, was preserved when appropriate proportions were maintained [44]:

The human body contains blood, phlegm, yellow bile and black bile. These are the things that make up its constitution and cause its pains and health. Health is primarily that state in which these constituent substances are in the correct proportion to each other, both in strength and quantity, and are well mixed.

The diversity, and also inconsistency, of thought is well illustrated in the *Timaeus*. First of all, like Hippocrates, Plato believes in a universe created by God [45], made up of four elements in important relation to one another:

. . . in the midst between fire and earth God set water and air, and having bestowed upon them so far as possible a like ratio towards another – air being to water as fire to air, and water being to earth as air to water, – he joined together and constructed a Heaven visible and tangible.

Later on, however, Plato mentions an unnamed fifth element, and refers to them all as compounds. By doing so, Plato alludes to combination rather than purity, a departure reinforced by suggesting there are different kinds of air including 'the most translucent kind which is called by the name of aether, and the most opaque which is mist and darkness' [46], as well as another form without a name. Nevertheless, the consistent belief in the balance of the elements required for health remains unwavering [47]:

The origin of disease is plain, of course, to everybody. For seeing that there are four elements of which the body is compacted, – earth, fire, water, and air, – when, contrary to nature, there occurs either an excess or deficiency of these elements, or a transference thereof from their native region to an alien region; or again, seeing that fire and the rest have each more than one variety, every time that the body admits an inappropriate variety, then these and all similar occurrences bring about internal disorders and disease.

What is clear from the first sections of this chapter is that air had a central place in the medical belief systems of many early civilisations. Even in the natural philosophy of the Greeks the conceptions of air reflect that health, disease and healing were understood as part of a bigger picture – humanity and the cosmos – and in terms of balance and harmony. This holistic, and often spiritual, way of thinking was to have an impact for many centuries to come.

Roman medicine

Although there were other important Roman contemporaries, the physician Galen stands out significantly for his influence in the history of medicine. His ideas built broadly on Hippocratic writings, so retaining the central place of air, but Galen synthesised Greek beliefs with his own insights in a way that proved long-lasting. A self-assured individual, Galen was happy to play to the crowds, and this arrogance and showmanship helped perpetuate his reputation [48]:

It is I, and I alone, who have revealed the true path of medicine. It must be admitted that Hippocrates already staked out this path . . . he prepared the way, but I have made it passable.

Galen was actually Greek, born to a wealthy family in Pergamon in Asia Minor (now Turkey) around AD 129. He had an extensive education and learned medicine from Alexandrian teachers, including visits to Egypt where he gained insights about treatments in India and Africa. He only arrived in Rome in AD 162, but soon gained a reputation and became physician to the powerful. He dissected animals extensively,[5] especially pigs and monkeys, and translated his findings to the human body – sometimes mistakenly. As physician to the gladiators, Galen gained some direct insights into human anatomy through examination of gaping wounds.

Galen wrote prodigiously, possibly 35 books, although few originals survived. Broadly speaking, Galen took the Hippocratic framework of health as balance, disease as imbalance, incorporated Platonic speculations on the macrocosm, and added a mixture of his own philosophical ideas and anatomical findings. As a physician, he treated people with concoctions of herbal and vegetable remedies (combined with heavy doses of confidence), causing some to label him as the first polypharmacist. Galen removed the Hippocratic emphasis on empiricism, on collating understanding of disease through observation of the ill, and focused instead on theorising and experimentation [10].

Because Galenic medicine drew so heavily on what went before, the significance of air remained, and can be found in three areas that have already been looked at in detail earlier. First, Galen held on to the Greek notion of balance of the four humours (black bile, yellow bile, phlegm and blood), which were in turn representations of the four elements – earth, air, fire and water. Next, Galen augmented pneumatism with his own ideas about circulation and anatomy. He believed that air, taken in through the lungs, combined in the heart with blood to generate *pneuma*, or vital spirit, the life breath of the cosmos. *Pneuma* was also modified in the liver (which formed blood from food) to create a natural spirit, which supported vegetative functions of growth and nutrition. From the heart, blood flowed to all organs, including the brain where a third alteration of *pneuma* resulted in animal spirit, distributed through the nerves to sustain movement and sensation, and without which animal life did not exist. Galen's system fitted in with Plato's divisions of the soul into the vegetative, animal and rational [49].

Last, Galen held on to the view that diseases were carried or transmitted by contaminated, polluted air: or miasma. This he defended in part on observational grounds, having witnessed occupational ailments, for instance slaves in copper mines who were obliged to make themselves masks from animal bladders as protection against the pungent, harmful atmosphere.

It was Galen, through his prolific output and social stature, who cleverly articulated Greek medical ideas, and the Roman empire that fostered their dissemination. But the legacy proved so enduring because Galenic medicine captured, in *pneuma*, the ingredient that allowed acceptance by (rising) Christianity of a vital spirit that could be considered close to the religious perception of the soul.

The Roman empire eventually collapsed, Europe descended into the Middle Ages, and it was not until emergence from the medieval period that Renaissance interest in science saw Galenic and Greek medical beliefs challenged. But new theoretical ideas did not really begin to appear until the advances in anatomical and physiological knowledge of the Enlightenment. And by the middle of the nineteenth century the place held by air in Western medical theory was changing.

NOTES

1. The second is *li*, the entity that organises all forms from above and the roots from which all things are produced. All beings, including humans, receive *li* in their moment of coming into existence and so obtain their specific nature.
2. *Shu* is what comes between *li* and *qi*, the way that the forces of nature operate.
3. Vitalism embodies the idea that the origin of life lies in a vital principle.
4. Three humours are depicted in Ayurvedic writings of ancient India: *vaya* (air); *pitta* (bile); and *kapha* (phlegm).
5. The dissection of humans, alive or dead, was not allowed in Roman society, although the Greeks sometimes permitted dissection on criminals – alive or dead.

REFERENCES

1. C. G. Helman, *Culture, Health and Illness: An introduction for health professionals*, 2nd edn (Oxford: Butterworth-Heinemann, 1990).
2. World Health Organisation, *World Health Organization (WHO) Constitution* (Geneva: WHO, 1946).
3. C. T. Onions, ed., *The Shorter Oxford English Dictionary on Historical Principles. Volume 1*, 3rd edn (Oxford: Clarendon Press, 1977), p. 565.
4. D. Greaves, *Mystery in Western Medicine* (Aldershot: Avebury, 1996).

5. British Medical Association Working Party, *Complementary Medicine: New approaches to good practice* (Oxford: Oxford University Press, 1993).

6. C. Bayley, Homeopathy. *J. Med. Phil.*, **18** (1993), 129–45.

7. R. Porter, *The Greatest Benefit to Mankind: A medical history of humanity from antiquity to the present* (London: Fontana, 1999).

8. P. Y. Ho and F. P. Lisowski, *A Brief Introduction to Chinese Medicine*, 2nd edn (Singapore: World Scientific, 1997).

9. *Ibid.*, p. 14.

10. F. F. Cartwright, *A Social History of Medicine* (New York: Longman, 1977).

11. J. Crace, *Quarantine* (London: Penguin, 1997), p. 25.

12. J. Longrigg, *Greek Medicine: From the Heroic to the Hellenistic age* (London: Duckworth, 1998).

13. J. R. Pinault, *Hippocratic Lives and Legends* (Leiden: E. J. Brill, 1992).

14. E. D. Phillips, *Greek Medicine* (London: Camelot Press, 1973).

15. G. E. R. Lloyd, ed., *Hippocratic Writings* (trans. J. Chadwick, W. N. Mann, I. M. Lonie and E. T. Withington) (London: Penguin, 1978), pp. 9–60.

16. Hippocrates, Breaths. In J. Longrigg, *Greek Medicine: From the Heroic to the Hellenistic age* (London: Duckworth, 1998), p. 45 (5).

17. —, The nature of man. In G. E. R. Lloyd, ed., *Hippocratic Writings* (trans. J. Chadwick, W. N. Mann, I. M. Lonie and E. T. Withington) (London: Penguin, 1978), p. 266 (9).

18. *Ibid.*, p. 267 (9).

19. —, Airs, waters, places. In G. E. R. Lloyd, ed., *Hippocratic Writings* (trans. J. Chadwick, W. N. Mann, I. M. Lonie and E. T. Withington) (London: Penguin, 1978), p. 148 (1).

20. —, Aphorisms. In G. E. R. Lloyd, ed., *Hippocratic Writings* (trans. J. Chadwick, W. N. Mann, I. M. Lonie and E. T. Withington) (London: Penguin, 1978), p. 213 (III.1).

21. *Ibid.*, p. 215 (III.19).

22. —, Airs, waters, places. In G. E. R. Lloyd, ed., *Hippocratic Writings* (trans. J. Chadwick, W. N. Mann, I. M. Lonie and E. T. Withington) (London: Penguin, 1978), p. 149 (2).

23. *Ibid.*, p. 150 (4).

24. —, Airs, waters, places. In G. E. R. Lloyd, ed., *Hippocratic Writings* (trans. J. Chadwick, W. N. Mann, I. M. Lonie and E. T. Withington) (London: Penguin, 1978), p. 151 (5).

25. —, Aphorisms. In G. E. R. Lloyd, ed., *Hippocratic Writings* (trans. J. Chadwick, W. N. Mann, I. M. Lonie and E. T. Withington) (London: Penguin, 1978), p. 213 (III.5).

26. —, Epidemics, Book 1. In G. E. R. Lloyd, ed., *Hippocratic Writings* (trans. J. Chadwick, W. N. Mann, I. M. Lonie and E. T. Withington) (London: Penguin, 1978), p. 94 (III.13/14).

27. —, Airs, waters, places. In G. E. R. Lloyd, ed., *Hippocratic Writings* (trans. J. Chadwick, W. N. Mann, I. M. Lonie and E. T. Withington) (London: Penguin, 1978), p. 156 (10).

28. *Ibid.*, p. 159 (11).

29. —, The sacred disease. In G. E. R. Lloyd, ed., *Hippocratic Writings* (trans. J. Chadwick, W. N. Mann, I. M. Lonie and E. T. Withington) (London: Penguin, 1978), p. 251 (21).

30. F. M. Cornford, *Plato's Cosmology* (London: Routledge and Kegan Paul, 1971).

31. G. Vlastos, *Plato's Universe* (Oxford: Clarendon Press, 1975).

32. A. E. Taylor, *A Commentary on Plato's Timaeus* (Oxford: Clarendon Press, 1972).

33. Plato, Timaeus. In *Plato: In twelve volumes (IX). Timaeus, Critias, Cleithophon, Menexenus, Epistles*, 7th edn (trans. R. G. Bury) (London: Heinemann, 1981).

34. T. A. Mahoney, Platonic ecology, deep ecology. In L. Westra and T. M. Robinson, eds., *The Greeks and the Environment* (Lanham: Rowman & Littlefield, 1997), pp. 45–54.

35. M. R. Adams, Environmental ethics in Plato's *Timaeus*. In L. Westra and T. M. Robinson, eds., *The Greeks and the Environment* (Lanham: Rowman & Littlefield, 1997), pp. 55–72.

36. Plato, Timaeus. In *Plato: In twelve volumes (IX). Timaeus, Critias, Cleithophon, Menexenus, Epistles*, 7th edn (trans. R. G. Bury) (London: Heinemann, 1981), p. 229 (85A).

37. *Ibid.*, p. 229 (85D–E).

38. *Ibid.*, p. 242 (7).

39. *Ibid.*, p. 243 (10).

40. Aristotle, On breath. In J. Barnes, ed., *The Complete Works of Aristotle: Volumes 1 and 2*, rev. Oxford edn (trans. J. F. Dobson) (Princeton: Princeton University Press, 1984), p. 767 (482b, 22–4).

41. *Ibid.*, pp. 768 (483a, 30).

42. R. Porter, *The Greatest Benefit to Mankind: A medical history of humanity from antiquity to the present* (London: Fontana, 1997), pp. 44–82.

43. F. B. Lund, *Greek Medicine* (New York: Paul Hoeber, 1936).

44. Hippocrates, The nature of man. In G. E. R. Lloyd, ed., *Hippocratic Writings* (trans. J. Chadwick, W. N. Mann, I. M. Lonie and E. T. Withington) (London: Penguin, 1978), p. 262 (4).

45. Plato, Timaeus. In *Plato: In twelve volumes (IX). Timaeus, Critias, Cleithophon, Menexenus, Epistles*, 7th edn (trans. R. G. Bury) (London: Heinemann, 1981), pp. 59–61 (32B).

46. *Ibid.*, p. 145 (58D).

47. *Ibid.*, p. 219 (82A).

48. Quoted in R. Porter, *The Greatest Benefit to Mankind: A medical history of humanity from antiquity to the present* (London: Fontana, 1997), p. 77.

49. J. Duffin, *History of Medicine: A scandalously short introduction*. Toronto: University of Toronto Press, 1999).

2

Miasma, contagion and survival of the fittest

In his wide-ranging review of the history of public health the historian George Rosen has suggested that not only did Hippocrates set the tone for 2000 years of medical thought, but the works of the Greek physician also provided the epidemiological reference upon which modern scientific medicine would be based [1]. Although the hagiography and historiography contained in Rosen's book now appear somewhat out of date and misplaced, there is – as the contemporary historian Elizabeth Fee comments in the introduction of a recent edition – value in Rosen's kind of overview, an effort which has perhaps not been superseded in public health history since its first publication in 1958 [2].

Notably Fee comments that public health history today needs to talk to practitioners in the field, to engage them in debates relevant to their work, and Rosen certainly tried to do just that. The links he draws between Greek medicine and the beginnings of public health provide a grounding framework for practitioners today, who can relate to the view that Hippocratic texts espoused the development of medical theory based on empirical observation, rationalism and notions of scientific induction.

In addition, the place of the environment (and air in particular) in Hippocratic thought about disease causation was firmly echoed in the period around the origins of public health in England in the mid-nineteenth century, and resonates strongly today. The following section from *Breaths* [3], for example, captures this broad Hippocratic association of polluted air with both specific complaints of individuals, and also with diseases affecting groups or communities:

There are two sorts of fever: the one which is common to all is called plague (loimos), the other, due to bad regimen, is specific and attacks those who follow a poor regimen. Air is the cause of both. The common (i.e. epidemic) fever has this characteristic because everyone inhales the same air, and when similar air is mingled in similar fashion with the body, similar fevers occur. But perhaps some will say, 'Why, then, do not such diseases attack all animals, but only a particular species?' I would reply that is because one body differs from another, one type of air from another . . . So, whenever the air has been defiled with such pollutions (*miasmasin*) which are hostile to human nature, then people fall sick . . .

This chapter explores the place of air in nineteenth-century medical thought in Britain. This leap of almost two millennia is not intended to diminish what happened in medicine in the intervening time, but is used to highlight significant developments in the conception of air and health around this important time in public health history. In particular air – objectified as a putative vehicle of disease – became the focus of debates about miasmatism and contagionism. Linked to this theoretical debate air, foul and dirty, became the object of environmental reform for the pioneers of public health.

Miasmatism and contagionism: the origins

According to Rosen, in the seventeenth and eighteenth centuries there were two conflicting conceptual explanations of epidemic disease causation: contagion or epidemic constitution. Ideas about contagion drew strongly on the works of the sixteenth-century Italian Girolamo Fracastoro, in particular his seminal book of 1546, *On Contagion, Contagious Diseases and their Treatment* [4]. In this Fracastoro argued that epidemic diseases were caused by transmissible and self-propagating minute infective agents, and these seeds, or *seminaria*, were specific for individual diseases.

Such particles were speculation until technologically revealed by the microscope, and first reported to the Royal Society in 1676 by the linen draper Anton van Leeuwenhoek who had observed wriggling creatures in soil, water and human excrement. However, mistakenly believing the observed particles (bacteria) to be spontaneously generated – the *product* of disease rather than the *cause* of it – meant contagionism did not dominate until the development of the germ theory of disease at the end of the nineteenth century [5].

In contrast, epidemic constitution, the alternative explanation of disease causation, held that epidemics were caused by the development of a state (or constitution) of the atmosphere, resulting from a constellation of weather conditions and local circumstances. This explanation drew strongly on Hippocratic ideas that local atmospheric conditions were at the root of diseases capable of spreading as long as the particular conditions lasted. Certain diseases were understood in relation to the broad environment and air was understood to be the mediator.

The famous seventeenth-century English clinician Thomas Sydenham, who first described the term 'epidemic constitution', divided febrile diseases into 'epidemic distempers' (e.g. smallpox) produced by atmospheric changes, and 'intercurrent diseases' (e.g. scarlet fever) which, although able to arise independent of the atmospheric state, were affected by it, but were also dependent on susceptibility of the body.

The influential Sydenham believed that the atmospheric change was due to a 'miasma' arising from the earth, and Rosen argues [6] that for most of the nineteenth century three theoretical positions on disease causation can be distinguished:

- miasmatic theory: epidemic outbreaks were caused by a state of the local atmosphere created by poor sanitary conditions;
- contagion: minute particles were the sole cause of infectious and epidemic diseases;
- limited or contingent contagionism: infections were caused by contagion, but only arose if other elements existed such as appropriate atmospheric conditions.

In contrast to Rosen's classical perspective on beliefs about disease causation in the nineteenth century, the historian Christopher Hamlin has postulated that the terms 'contagion' and 'miasm' belonged to a larger and more complex system of causation. The distinction between contagion (the vehicle of person-to-person disease transmission, only received from a previous human host) and miasm (pathogenic emanations dispersed into the atmosphere in which disease could spontaneously generate) was not always accepted. Sometimes the terms were used synonymously, sometimes not; sometimes they were used to answer different questions [7].

Hamlin argues convincingly that, although the terms implied disease specificity, they were in fact vaguely and variously used and might be among many malignant forces felt to harm bodily constitution. What was clear, however, was that with both contagion and miasm disease reached victims through the air. Although distance from the source differed with each term, air was the medium of disease.

So, two thousand years after the Hippocratic texts were written, the theory of miasmatism apparently remained part of at least two broad, competing understandings of disease causation. The etymological roots of such an important word are revealing. Stemming from the Greek *Mia* word group – whose basic meaning is that of defilement or impairment of a thing's form or integrity – miasma essentially refers to pollution or impurity.

The impurity, however, could relate to something physical or moral, and in Greek times the two were often interwoven. Parker, for instance, suggests that in classical antiquity the word 'miasma' could equally have been used to refer to a form of communicable religious danger, the gods seeming irrelevant, or a dangerous dirtiness that individuals rub off one another like a physical taint [8].

This multiplicity of meaning remained highly significant in the nineteenth century. The polluted air felt to be, in one way or another, causally related to infectious diseases, was held by some to result from the immoral behaviour of the poor. In miasma, physical and moral pollution were bound together linguistically and metaphysically and, as described in the next section, the debate about infectious disease causation was shaped by the developing theory of evolution by natural selection with its emphasis on the environment. Later, the extension of this theory into the realm of moral evolution occurred simultaneously with the establishment of public health in Britain.

Air and the origins of public health

A somewhat traditional notion of public health purports Edwin Chadwick as the founder, and places the origins in the first half of the nineteenth century. However, as Dorothy Porter points out, public health – defined by her as collective action in relation to the health of populations – had been going on for centuries but only really acquired a professional and institutional foundation in the middle of the nineteenth century [9].

Taking this distinction into account, air was at the heart of the traditional notion of the birth of (professional and institutional) public health in England in two inter-related ways. First, the debate about disease causation was central because it underpinned efforts made to improve conditions and thereby reduce diseases, which formed the mainstay of early public health efforts. Second, the place of the environment, of which air was a crucial component, in directing human progress was being explored. Inextricably linked to these was the issue of responsibilities: individual responsibility for creation of the conditions in which one lives; and State responsibility for improving the living and working conditions of its citizens.

Considered by some as the founder of public health, the barrister Edwin Chadwick was first notable as the architect and enforcer of the unpopular new Poor Law of 1834, a law designed to make the conditions under which public relief could be guaranteed so unpleasant that most would refuse to request it. Believing that easy charity contributed to, or even created, indigence, the law sought to focus on prevention, which was felt to be cheaper than relief [10].

Chadwick has been described as an inductive social scientist, who built up data on mortality, diet and the environment to test generalisations that diseases caused by environmental filth engendered destitution. These investigations led to publication of his famous 1842 *Report on the Sanitary Conditions of the Labouring Population of Great Britain*, in which he argued that insanitary conditions led to social, biological and psychological problems and, by inference, good sanitation should lead to a happy, healthy proletariat [11].

Like others, Chadwick believed that local atmospheric conditions were responsible for certain infectious diseases affecting communities. Processes such as putrefaction at ground level, combined with poor urban sanitation and drainage, created a residue of filth that contaminated the air, and these local atmospheric states caused disease. Believing cause and cure to be linked, Chadwick labelled 'atmospheric impurity, occasioned by means within the control of legislation, as the main cause of the range of endemic and contagious diseases among the community, and as aggravating most other diseases' [12].

As well as identifying the single main cause of ill health, Chadwick specified that control of insanitary conditions was within the State's remit and jurisdiction. The Public Health Act of 1848, which followed from Chadwick's *Report*, put this control in motion by providing local boards of health with legislative powers and money to improve local sewage and sanitation, which would in turn improve atmospheric conditions. Although often invoked as marking the birth of public health in Britain, this somewhat compromising bill, which failed to cover smoke prevention or insanitary burial, did not really develop teeth until updated as the 1875 Public Health Act, which finally curbed much of what the previous bill had pronounced as permissive.

Yet underscoring both of these bills lay the reasons for the desired improvement in living and atmospheric conditions. Chadwick may have engendered the notion that people's health was a matter of social concern, but this position was grounded in the perceived need to have a healthy workforce in a time of rapid industrialisation and economic development. There certainly was, as Hamlin has described [13], a utopianism to the early public health efforts of Chadwick, but this was founded on the utilitarian requirements of an expanding Empire. As is well known, Chadwick was Jeremy Bentham's secretary and follower in earlier years.[1]

Where air fitted in was in bringing together the need for healthy workers with the scientific explanation of ill health. If insanitary environmental conditions caused atmospheric impurity, and atmospheric impurity was responsible for disease, then improving the health of the poor should result from better environmental conditions. However, though framed in terms of social justice and welfare, the movement was not driven by the same passion and egalitarianism which motivated others such as Engels, at a similar time, carefully to observe and document the association between working conditions and disease [14].

Instead, the movement made sense politically and economically, and fitted into the growing belief in the scientific explanation of human progress driven by the environment. If the environment directed evolution in the animal kingdom, then it made sense that the environment could induce debility in human beings, and if the character of the poor, considered morally inferior by some, could not be trusted to improve conditions, then the State needed to act.

Here, however, the explanation provided by environmental determinism seemed to be at odds with the action advocated by believers in social evolution. Therefore, to understand better how the place of air in the scientific debate about infectious disease causation fitted in with public health action recommended to improve the air, it is essential to look more closely at the inextricably linked debate going on about the place of the environment in directing human progress.

Air, the environment and evolution

The development of Darwin's theory

Although Charles Darwin did not publish *The Origin of Species by Means of Natural Selection* until 1859 [15], it is well established that he had developed the main tenets of his theory at least two decades earlier. His transmutation notebooks [16] – written shortly after returning from the famous voyage on the *Beagle*[2] – indicate his early reasoning,[3] and in 1842 he wrote a 35-page *Sketch* of the theory which was developed into a longer *Essay* [17] two years later. Darwin left his wife to publish this essay in the event of his death [18]. It was not, however, until 1858, prompted by a letter from Alfred Russel Wallace[4] and urged by his close friends Charles Lyell and Thomas Huxley, that Darwin openly revealed his ideas by reading a joint paper with Wallace to the Linnaean Society [19]. The following year his classic work was published.

A number of reasons for Darwin's almost two-decade-long delay have been put forward: psychological explanations based on Darwin's timid character and his concern about the impact of his theory [20, 21]; scientific explanations centred around the time Darwin needed to gather further evidence, necessitating the eight-year study of Cirrepedes, a subclass of barnacles [22, 23], and religious explanations exploring the changing nature of Darwin's theism and developments in natural theology, which made the religious climate more accepting of evolutionary ideas [24–26]. It is most likely that all these interconnected internal and external factors (the man himself and the social context changing considerably over a 20-year period) contributed to the delay, progressing towards an atmosphere in which his theory was more palatable [27].

Of significance to this chapter is that, during the whole period around 'Darwin's delay' there was a growing debate about mankind's place in nature, a debate which allowed for the articulation and acceptance of Chadwick's beliefs about disease causation and the remedial action required. Although fuelled by scientific developments, the debate, of which Darwin was one of many central figures, was largely sociopolitical and was manipulated for a variety of purposes.

Before looking at these purposes, it is worth recapping the basic tenets of evolutionary theory. Building on the ideas of the earlier writers, Darwin's theory of evolution by natural selection described the environment as being the space that directed development of a species. The environment was understood in a broad physical sense including geography, climate and predation. Individual members of a species best adapted to the environment were most likely to survive and procreate. This concept of 'relative adaptation', combined with competition between individuals for scarce resources, resulted in survival of the fittest members. In turn, this was most likely to lead to the flourishing and perpetuation of the species, which was considered the evolutionary purpose. With genetic inheritance not understood for some decades to come, Darwin was unaware of the mechanisms by which adaptation was passed on to successive generations.

The social significance of evolutionary theory

Apart from the biological and scientific importance of these new ideas, evolutionary theory held huge social significance. Although the theory was formulated mainly from research on the non-organic and organic non-human worlds, Darwin was interested in universality from early on. Although he refrained from discussing humans at length until much later, in the summary of *Origin of Species* Darwin gave an early indication that the operation of natural law could be extended to all organic beings [28], presumably including mankind:

In the survival of favoured individuals and races, during the constantly-recurrent Struggle for Existence, we see a powerful and ever-acting form of Selection. The struggle for existence inevitably follows from the high geometrical ratio of increase which is common to all organic beings.

If mankind was included, the question faced was to which human characteristics would such laws apply, and to what ends? On a simple, physical level, extension to humans could explain different physical characteristics of individuals within different races, determined by human evolutionary adaptation to different surroundings. Darwin himself noticed this early on, commenting on the physical attributes of native islanders on his sea-faring journey. Taken further it was not difficult to see how, within a given race of humans, the strongest individuals might be most likely to survive, whether in the harsh undeveloped surroundings of distant islands or elsewhere [29].

However, evolutionary theory held the greatest impact when applied to human beings at a level higher than the individual. Although individuals undoubtedly competed with each other, if evolution acted within human societies at the group or race level, then success or failure of groups or races could be explained, and more importantly legitimated, as being in the best – and natural – interests of the species, mankind. Although Darwin, late in life, did indeed defend imperialist success as evidence that natural selection had done much for the progress of civilisation through elimination of the lower races, such beliefs were widespread much earlier that century [30].

One of the foremost proponents of such views was Herbert Spencer, a philosopher regarded as one of the founders of sociology, best known for his ten-volume *System of Synthetic Philosophy*, a work spanning 40 years beginning with the 1862 *First Principles* [31]. Within this work Spencer attempted to show how his Principle of Evolution is exemplified throughout organic and inorganic nature, including the individual, social and moral life of humanity. Spencer argued that the evolutionary direction of flow of events is from simple to complex, incoherent to coherent, undifferentiated to differentiated, homogeneous to heterogeneous, uniform to multiform. Spencer's ideas are evident in his earlier writings in the 1840s and 1850s, and the belief that everything is under the influence of an evolutionary force is sharply illustrated in an essay of 1857 [32]:

... this law of progress is the law of all progress. Whether it be in the development of the Earth, in the development of the Life upon its surface, in the development of Society, of Government, of Manufacturers, of Commerce, of Language, Literature, Science, Art, the same evolution of the simple into the complex, through successive differentiations holds throughout.

Spencer's deterministic theory was based upon a complete belief in universal and inevitable development to perfection, which included mankind and morality. Society passes through necessary stages, of which the industrial was simply the present, en route to the social state which – corresponding to the transition from egoism to altruism – Spencer called the 'end-product of history' [33]. At this point the ideal man, exhibiting perfect morality, would live in the ultimate cooperative society where no evil existed.

Although competition between societies was natural and for an ultimate reason, Spencer felt that nations acting with malevolent force would not flourish long-term. He did not believe in the existence of Bentham's transcendental unitary human nature, instead supporting a sort of moral and political relativism, all explained as part of the path to perfect adaptation. But he disagreed vehemently with any State intervention, since it was interference with the natural process, believing there 'cannot be more good done than of letting social progress go on unhindered' [34].

Spencer's ideas – influential from the Athenaeum to the USA [35] – held great appeal to industrialists, part of the rising professional middle class who wanted to reap the benefits of their own work. With unlimited competition considered natural, and State interference unwarranted, Spencer provided the perfect justification for both the behaviour of the capitalist and *laissez-faire* opposition to State-induced reform. Several decades earlier Adam Smith had also espoused free trade and an open market, as well as subtly connecting how the hidden hand of economics shapes political arrangements, our values and ultimately how people behave [36]. Supporters of Jeremy Bentham drew selectively on Spencer's philosophy and argued that the evolutionary view of nature's fierce struggle for survival legitimised the belief in free market social competition and the abolition of monopolies voiced by Scottish political economists.

Evolutionary ideas, interestingly, did not only appeal to the right wing. In *The Politics of Evolution*, the historian Adrian Desmond shows how in the 1820s and 1830s there was an influx of medically trained teachers from Paris and Edinburgh to London, espousing the evolutionary ideas of earlier writers such as Lamarck [37]. Arguing for a levelling of nature, the medical proponents of radical views, led by Darwin's Edinburgh mentor Grant,[5] were kept at a distance by the profession's establishment, especially the Royal College of Surgeons. They created an opposition campaign, which was part of the secularising, dissenting agitation, aimed at redistributing medical privilege, and democratising the institutions of science and State.

At the same time, working-class radicals tied Lamarckian evolution to socialist strategies and atheist demands as they tried to wrest power from Tory Anglicans. Materialism and transmutation were thrust from early on into a confrontationalist class context [38]. A little later, focusing on socialists and their connection to Darwin, Himmelfarb quotes Karl Marx suggesting in 1861 after reading *Origin of Species*, that the book provides 'a basis in natural science for the class struggle in history' [39]. Marx valued the book as he felt it offered a death-blow to teleology in the natural sciences, and moved away from studies which tended to exclude history and its processes. Engels [40], however, heralded not Darwin, but Marx as the foremost social evolutionist:

Just as Darwin discovered the law of evolution in organic nature, so Marx discovered the law of evolution in human history.

Moral evolution, the environment and public health

As strong as the biological arguments upholding the legitimacy of human social evolution were, for those with political leanings toward both extremes the concept was necessarily tied in with developing ideas about moral evolution. For industrialists and expansionists in particular there was a problem since, although evolutionary theory placed humans squarely within the same laws of nature as other species, surely humans remained different by virtue of cognition and morality. And some felt that, despite scientific support, it still did not seem right to behave aggressively and unsympathetically towards other humans.

This problem could be circumvented if one accepted mind as matter, with human mental characteristics subject to the same evolutionary laws as physical characteristics. It was then possible to postulate a hierarchy of social morality grounded in degree of social development. Evolutionary forces shaped moral progress with the most superior or successful races having the highest degree of morality. Darwin was himself a believer in philosophical materialism and the inheritability of acquired psychological traits. By 1871 he had developed ideas that complex social instincts in animals, resulting from natural selection acting on simple instincts, provided the foundation for evolution of 'the most noble of all the attributes of man' [41, 42], morality:

... any animal whatever, endowed with well-marked social instincts, would inevitably acquire a moral sense of conscience, as soon as its intellectual powers had become as well developed, or nearly as well developed as in man.

Progression to morality within nature continued within the groups and races of mankind. Near the beginning of *Descent of Man* Darwin set out a scale in which 'civilized man' appears at the top, then natives, followed by barbarians, savages and apes, lending additional support to the morally questionable aggressiveness required for colonial expansion [43]. Closer to home the 'careless, squalid, unaspiring Irishman ... who multiplies like rabbits' [44] exemplified how it became possible to cast the worse-off groups in society as being in that position by virtue of their morally inferior behaviour. So the environment weeded out not only the most unfit physically, but also the most unfit morally. In this way groups such as the poor and destitute could be blamed, at least in part, for their situation.

In this milieu of interwoven scientific and social thought, the beliefs of Chadwick and others that the environment – and air in particular – determined health and disease, life and death, made sense. Whatever the strength of scientific support of these medical views (which the germ theory of disease was to supersede some decades later), the idea that adverse environmental conditions were causally responsible for disease which might weed out those exposed to such conditions, the unfit, was commensurate with the ideas of human social evolution.

These medical beliefs were lent further weight by the new perspectives on moral evolution. By virtue of their morally inferior behaviour the unfit were, at least in part, responsible for the conditions which caused their ill health. It was therefore not surprising to find comments such as those in a *British Medical Journal* editorial of 1870 [45], connecting morality and social position, and placing partial blame for their own welfare with those who suffered most directly from the bad air and poor sanitation:

To members of the medical profession it must be well known, how intimately overcrowding, dirt, and low moral condition, are connected. Medical officers of health, and indeed all who are engaged in the treatment and prevention of disease, meet with great hindrance through the ignorance or carelessness of those with whom they have to deal. They may succeed in enlightening people as to the evil of this or that habit ...

Of course, the sanitary reform advocated by Chadwick and his followers – support of the poor via improvement to environmental conditions – seemed to be working against the forces of nature. Increasing the chances of survival of the weakest members of the race would appear to run counter to the dictates of biological determinism. Once again Darwin's voice reflected widely held views of the time [46]:

We civilized man . . . do our utmost to check the process of elimination; we build asylums for the imbecile, the maimed, and the sick; we institute poor laws; and our medical men exert their utmost skill to save the life of every one to the last moment . . . thus the weak members of civilized societies propagate their kind . . . this must be highly injurious to the race of man.

But it was here that the early public health efforts could be defended from a higher level of social and moral progress. The reforms served to improve the overall condition of the working nation, which was essential to economic growth and further colonial development. In the long-term the human species would be the beneficiary of the continued expansion and domination of the British race, believed to represent the finest physical and moral characteristics of the species.

Further evidence was provided by upholders of scientific racism (part of the doctrine of biological determinism),[6] who tried to demonstrate scientifically the inborn inferiority of some races relative to others [47]. Observed racial differences in body shape, brain size [48] and brain complexity [49], provided empirical justification for actions explained by evolutionary superiority, and laid the foundations for the most extreme versions of Social Darwinism to follow – eugenics (defended as giving nature a helping hand) – and National Socialism [29].

Michel Foucault ties together the apparent ideological conflict between those supporting and those opposing support of the poor through environmental improvement with his concept of 'biopower' – the controlled insertion of bodies into the machinery of production and the adjustment of the phenomenon of 'population' to economic processes [50]. He sees in eugenics one example of population control as a political practice, which began during the eighteenth and nineteenth centuries with progressive dismantling of the means of assistance to unfortunate groups, such as the 'sick poor'. During industrialisation the workforce needs to be carefully managed, and biopower, indispensable to capitalism, involved redefining the poor according to their economic usefulness, placing, for example, the 'healthy poor' into the circuit of production [51]. So, in this sense, environmental and sanitary reform could be seen as a necessary part of sustaining a healthy poor workforce, with any benefit to the sick poor an accepted by-product.

This digression into the evolutionary debate is not intended to paint a purely calculating and bleak picture of the origins of public health in nineteenth-century Britain. As pointed out earlier there was a certain degree of utopianism, and a feel for social justice in the ideals of those involved. But these ideals were allowed to take root in practice because understanding the environment as a substantial explanation of human well-being fitted in with corresponding biological theories of the environment as the determinant of human social and moral progress, as well as the economic and political dictates of the time.

So the conception and place of air in medical thought had changed dramatically. From being an essential, inseparable part of holistic understanding of human health and disease, air had become the medium around which a scientific debate focused, a debate which itself was framed within a larger debate about the environment and human progress.

NOTES

1. Bentham, one of the founding figures of utilitarianism, is discussed further in Part IV.
2. Darwin's *Notebooks on the Transmutation of Species* were written between July 1837 and July 1839.
3. Thomas Malthus wrote two repeatedly revised *Essays on Population* (1798 and 1803), read by Darwin when he was formulating his theory and likely to have influenced him.
4. In this letter Wallace outlined a theory similar to Darwin's.
5. Grant first introduced Darwin to the idea of transmutation.
6. Stephen Jay Gould describes biological determinism as the doctrine stating that shared behavioural norms, and the social and economic differences between human groups – primarily races, classes and sexes – arise from inherited, inborn distinctions, and that society, in this sense, is an accurate reflection of biology [29].

REFERENCES

1. G. Rosen, *A History of Public Health*, expanded edn (Baltimore: Johns Hopkins University Press, 1993).
2. E. Fee, Public health, past and present: a shared social vision. In G. Rosen, *A History of Public Health*, expanded edn (Baltimore: Johns Hopkins University Press, 1993), pp. ix–lxvii.
3. Hippocrates, Breaths (6). In J. Longrigg, *Greek Medicine: From the Heroic to the Hellenistic age* (London: Duckworth, 1998), pp. 124–5.
4. G. Fracastoro, *On Contagion, Contagious Diseases and Their Treatment* (trans. W. C. Wright) (New York: Putnam, 1930).
5. G. Rosen, *A History of Public Health*, expanded edn (Baltimore: Johns Hopkins University Press, 1993), pp. 79–83.
6. *Ibid.*, pp. 263–6.
7. C. Hamlin, *Public Health and Social Justice in the Age of Chadwick: Britain, 1800–1854* (Cambridge: Cambridge University Press, 1997).
8. R. Parker, *Miasma: Pollution and purification in early Greek religion* (New York: Oxford University Press, 1983).

9. D. Porter, *Health, Civilisation and the State: A history of public health from ancient to modern times* (London: Routledge, 1999), pp. 1–8.

10. C. Hamlin and S. Sheard, Revolutions in public health: 1848, and 1998? *BMJ*, **317** (1998), 587–91.

11. C. Hamlin, State medicine in Great Britain. In D. Porter, ed., *The History of Public Health and the Modern State* (Amsterdam: Editions Rodopi, 1994), pp. 132–64.

12. Quoted in C. Hamlin, *Public Health and Social Justice in the Age of Chadwick: Britain, 1800–1854* (Cambridge: Cambridge University Press, 1997), p. 160.

13. C. Hamlin, State medicine in Great Britain. In D. Porter, ed., *The History of Public Health and the Modern State* (Amsterdam: Editions Rodopi, 1994), p. 146.

14. F. Engels, *The Condition of the Working Class in England in 1844* (Moscow: Progress Publishers, 1973). [First published in German in 1844. First published in English in 1888.]

15. C. Darwin, *The Origin of Species by Means of Natural Selection, or the Preservation of Favoured Races in the Struggle for Life* (New York: New American Library, 1958). [Originally published in 1859.]

16. G. de Beer, ed., *Autobiography of Charles Darwin* (Oxford: Oxford University Press, 1974).

17. —, ed., *Evolution by Natural Selection* (Cambridge: Cambridge University Press, 1958).

18. P. Bowler, *Charles Darwin: The man and his influence* (Oxford: Blackwell, 1990).

19. C. Darwin and A. Wallace, On the tendency of species to form varieties, and on the perpetuation of varieties and species by means of natural selection. *J. Proc. Linn. Soc., Zoology.* **3** (1858), 45–22. http://pages.britishlibrary.net/charles.darwin3/tendency1858.html (accessed 10 February 2005).

20. R. Colp, *To Be an Invalid* (Chicago: University of Chicago Press, 1977).

21. A. La Vergata, Images of Darwin: a historiographic overview. In D. Kohn, ed., *The Darwinian Heritage* (Princeton: Princeton University Press, 1982), pp. 901–72.

22. M. Ghiselin, *The Triumph of the Darwinian Method* (Berkeley, University of California Press, 1969).

23. D. Ospovat, *The Development of Darwin's Theory* (Cambridge, Cambridge University Press, 1981).

24. A. Desmond, *Archetypes and Ancestors* (London: Blond and Briggs, 1982).

25. J. Brown, The evolution of Darwin's theism. *J. Hist. Biol.*, **19** (1986), 31–45.

26. S. Schweber, The origin of the *Origin* revisited. *J. Hist. Biol.*, **10** (1977), 310–15.

27. R. Richards, Why Darwin delayed, or interesting models in the history of science. *J. Hist. Behav. Sci.*, **19** (1983), 45–53.

28. C. Darwin, *The Origin of Species by Means of Natural Selection, or the Preservation of Favoured Races in the Struggle for Life* (New York: New American Library, 1958), p. 400. [Originally published in 1859.]

29. S. J. Gould, *Ever Since Darwin* (London: Burnett Books, 1977).

30. J. Greene, Darwin as a social evolutionist. *J. Hist. Biol.*, **10** (1977), 1–27.

31. H. Spencer, *First Principles* (London: Williams and Norgate, 1864). [Originally published in 1862.]

32. —, Progress: its law and cause. In H. Spencer, *Spencer's Essays: A selection* (London: Watts and Co., 1907), p. 8. [Originally published in 1857.]

33. —, *Social Statistics or the Conditions Essential to Human Happiness Specified, and the First of Them Developed* (London: Williams and Norgate, 1850), p. 419.

34. —, *The Study of Sociology* (London: Williams and Norgate, 1873), pp. 401–2.

35. R. Hofstadter, *Social Darwinism in American Thought* (Boston: Beacon Press, 1944).

36. A. Smith, *The Wealth of Nations*, 6th edn (London: Methuen, 1950). [Originally published in 1776.]

37. A. Desmond, *The Politics of Evolution* (Chicago: University of Chicago Press, 1989).

38. R. Young, Darwinism *is* social. In D. Kohn, ed., *The Darwinian Heritage* (Princeton: Princeton University Press, 1982), p. 609–38.

39. G. Himmelfarb, *Darwin and the Darwinian Revolution* (London: Chatto and Windus, 1959), p. 347.

40. Quoted in G. Himmelfarb, *Darwin and the Darwinian Revolution* (London: Chatto and Windus, 1959), p. 349.

41. C. Darwin, *The Descent of Man and Selection in Relation to Sex* (two volumes) (London: John Murray, 1871), p. 70.

42. *Ibid.*, p. 71.

43. *Ibid.*, p. 35.

44. *Ibid.*, p. 172.

45. Anonymous, The public health problem. *BMJ*, 1 October (1870), 361–2.

46. C. Darwin, *The Descent of Man and Selection in Relation to Sex* (two volumes) (London: John Murray, 1871), p. 168.

47. M. Harris, *The Rise of Anthropological Theory* (London: Routledge and Kegan Paul, 1968).

48. C. Morton, *Crania Americana* (Philadelphia: John Pennington, 1839).

49. E. Tylor, *Primitive Culture: Researches into the development of mythology, philosophy, religion, language, art and custom* (London: John Murray, 1871).

50. M. Foucault, Right of death and power over life. In M. Foucault, *The History of Sexuality. Volume 1. An Introduction* (New York: Random House, 1978).

51. —, The politics of health in the eighteenth century. In C. Gordon, ed., *Power/knowledge: Selected interviews and other writings 1972–1977* (Brighton: Harvester Press, 1980), pp. 166–83.

Conclusions to Part I

A historical overview has been presented in these two chapters, in an attempt to explore the theme of air and health over two millennia. In ancient civilisations – Egyptian, Chinese, and Judeo-Christian – air was a spiritual essence in the conceptualisation of human health and disease, a provider and sustainer of life. This theme continued in Greek medicine, within which the conception of air and health had both natural and supernatural dimensions, and was associated with a holistic conception of health and disease, linked to balance and harmony with nature and the universe.

These conceptions were largely lost with the advent of scientific medicine in the West. Air became the object of scientific deliberation in nineteenth-century Britain, and its role in disease causation became linked to wider questions about human evolution and progress, as well as the economic and political requirements of the period. The next part of this book picks up chronologically where this part ends but looks at a different conception – polluted air and its scientific examination – and traces this into the twentieth century.

Polluted Air

Overview of Part II

In Part I it was seen how early conceptions of air and health retained a sense of spiritualism and holism. But, by the time of the birth of public health in Britain, air had become the medium around which a critical scientific debate about disease causation was taking place, this debate itself framed by a wider discussion about how the environment directed human social and moral development. While holding on to a sense of 'environmentalism' a narrower conception of air was beginning to appear, one which focused more specifically on air as part of the broader physical environment, and how air affected human health and disease.

After a short step back, Part II traces the changing conception of air in medicine from the mid-nineteenth century until about 1970. As Western science began to underpin medical theory and practice during the second half of the nineteenth century, air became predominantly conceptualised as polluted air, and those involved in public health began to search for scientific evidence of links between bad air and ill health. Such evidence was not hard to demonstrate, but policy to improve the air failed to follow.

It is argued in Part II that this relationship between air pollution science and air pollution policy has continued through the twentieth century. The effects on health of polluted air have long been recognised and repeatedly illustrated, albeit with certain scientific limitations. Further, polluted air has been gradually re-conceptualised as its constituent air pollutant components, and air pollution epidemiology has correspondingly tended towards reductionism in exposure and outcome measurement.

But legislation to reduce air pollution levels and attenuate the adverse effects on health has generally either failed to appear or failed to deliver. The story of procrastination and undelivered promises has continued despite the relentless campaigning of pressure groups and, in the first half of the twentieth century, by public health professionals. Changes, however, in the structure and organisation of public health have distanced the public health worker from research, advocacy and 'ground-level' action.

Developments in air pollution policy have mainly been dictated by economics and politics. Reducing pollution costs money and requires lifestyle changes generally felt to be politically unpalatable. What successes there have been can be attributed to a combination of concerns about visibility, aesthetics, washing and falling prices of smokeless fuels.

The problems of polluted air outlined here relate to developments in an industrialising Western country. However, on a worldwide scale today, the ill health resulting from the indoor use of wood-burning stoves in low-income countries remains a more significant global public health problem than outdoor air pollution, despite the greater attention the latter receives. This is looked at in Chapter 6.

3

As far as is practicable: air pollution policy and public health in Britain 1800–1900

For reference, a historical overview of smoke and air pollution legislation in England is provided in Table 3.1.

Early efforts to control smoke

While the middle of the nineteenth century marked a reorientation from the integrated place of air in medical systems of thought to the systematic focus on the health effects of polluted air, the roots of these health concerns can be traced back much further. In fact, as soon as humans began to use natural materials to create heat, the side-effects of the burning process would have been recognised.

The first raw material used was wood, the burning of which creates dense smoke and particles which are irritants to the respiratory tract, especially when used indoors: there is evidence of irritant-induced sinusitis and blackening of lung tissues in Anglo-Saxon times [1–3]. However, the polluted air that marked the conceptual shift described above was not polluted with the by-products of wood burning,[1] but instead with the smoke resulting from the burning of coal for heat, cooking, and later steam, in homes and factories.

Although it is possible that coal was first used in the UK in Glamorganshire, Wales, during the Early Bronze Age – around 1500 BC [4] – it was probably first discovered in England in the ninth century. As populations grew, along with developments in early manufacturing methods, wood supplies in afforested areas started dwindling in Europe and coal was turned to more systematically as a substitute [5].

Possibly the earliest concerns about the effects of coal are the smoke-induced repairs to Nottingham Castle that Henry III directed in the 1250s, and an Ordinance of 1273 which prohibited the use of coal in London as being prejudicial to health – the authenticity of the latter has been questioned [6]. By the end of the thirteenth century an appreciable quantity of coal was being used in London, and investigative commissions were convened in 1285 and 1288 to look into the problem, a political tactic that has endured through the history of smoke pollution.

Table 3.1 Overview of smoke and air pollution legislation in England 1273–1968

Year	Legislation	Details
1273	Ordinance	Ordinance prohibiting use of coal in London as being prejudicial to health.
1306	Royal Proclamation	Issued by Edward I forbidding the use of coal by artificers, who were to return to charcoal.
1388	English Sanitary Act	First English Sanitary Act, dealt with offal and slaughterhouses, and prohibited casting of animal filth and refuse into rivers or ditches, and corrupting of the air.
1467	Local law	Law passed in Yorkshire forbidding the building of any more kilns because of the stink and badness of the air, and detriment to fruit trees.
1845	Railways Clauses Consolidation Act	Required railway locomotives to consume, as far as practicable, their own smoke.
1847	Town Improvement Clauses Act	Required factory furnaces to consume, as far as practicable, their own smoke.
1853	Smoke Nuisance Abatement (Metropolis) Act	Empowered Home Office to appoint an inspector, working in consultation with metropolitan police, to abate nuisance from the smoke of furnaces in the metropolis and from steam vessels above London Bridge.
1863	Alkali, etc., Works Regulation Act	Empowered appointment of inspectors to inspect air pollution from certain factories.
1875	Public Health Act	Legislated that any fireplace or furnace in trade premises must as far as practicable consume its own smoke, and enabled action to be taken against those responsible for factory chimneys emitting black smoke in sufficient quantity to be a nuisance. The Act did not cover private dwelling-houses and did not apply to London.
1891	Public Health (London) Act	Conferred similar provisions against smoke nuisance to London as contained within the 1875 Act.
1907	Public Health (Amendment Act)	Empowered local authorities to make by-laws regarding construction of chimney shafts for furnaces of steam engines and certain factories.
1926	Public Health (Smoke Abatement) Act	No longer necessary to prove smoke was black when emitted in sufficient quantity to be a nuisance; extended definition of smoke to include soot, ash, grit and gritty particles; enabled local authorities to make by-laws to control emission of smoke.
1956	Clean Air Act	Control of smoke emissions from factories; introduction of smoke-control areas.
1968	Clean Air Act	Revised the Clean Air Act 1956, and extended it to prohibiting emission of dark smoke from industrial and trade premises.

The effects of coal burning were especially noticeable to those entering London, and noble lords coming to the metropolis to attend Parliament found the smoke objectionable. Their complaint to Edward I resulted in the king issuing a royal proclamation in 1306 forbidding the use of coal by artificers, who were to return to charcoal. Though one offender was apparently executed [7], the proclamation proved largely ineffectual, and the following year a Committee of Inquiry was appointed with instructions to 'inquire of all such who burnt sea-coal in the city or parts adjoining, and to punish them for the first offence with great fines and ransoms, and upon the second offence to demolish their furnaces' [8]. But rising wood prices continued to make coal an attractive alternative.

At the time of the reign of Queen Elizabeth I (1558–1603), coal production had grown to approximately 50 000 tons a year, which was largely industrial in use and attributable to lime production in particular. Domestic coal use remained less popular, except for the poor, because of the degree of smoke produced without some removal arrangement [6]. The queen forbade the use of coal when Parliament was in session, complaining that smoke from breweries in the vicinity of her palace caused her grievous annoyance. Early policy measures such as these, arising from noble or royal discontent, reflect that action is more likely to result when those in power are directly bothered by the problem in question. Seeds of the modern environmentalist term NIMBY (Not In My Back Yard) can be recognised – a dig at middle-class lack of interest in ecological problems unless they are on one's doorstep.

During the seventeenth century the amount of coal imported to London increased substantially. Concern about the pall of coal smoke over cities, particularly London, had increased, with wealthier individuals and families already choosing to reside on the fringes of the capital. In 1648 Londoners petitioned Parliament to prohibit the importation into London of coal from Newcastle [9].

In 1661 the diarist John Evelyn (1641–1706) published a famous tract entitled *Fumifugium, or the Inconvenience of Aer and the Smoake of London Dissipated*, which he dedicated to Charles II [10]. In this vivid piece Evelyn drew a dramatic picture of London darkened and eclipsed by a hellish and dismal cloud of sea coal, belched forth from the sooty jaws of the tunnels of brewers, dyers, lime burners and soap boilers. Presaging early twentieth-century smoke-control advocates' focus on aesthetics and cleanliness, Evelyn powerfully depicted the dirtiness of 'this horrid Smoake which obscures our Churches, and makes our Palaces look old, which fouls our Clothes, and corrupts the Waters . . . diffuses and spreads a Yellownesse upon our choycest Pictures and Hangings . . . and kills our *Bees* and *Flowers*' [11].

Evelyn was not working alone, and his work is significant as the product of careful observation and documentation of the effects of pollution (by himself and others), influenced by the thinking of Francis Bacon at a time of rapid progress in science [6]. Evelyn's tract is also important because it highlighted the perceived adverse health effects of prolonged exposure to foul air, and also illustrated a simple method (supported by statistics) – to be used often from the nineteenth century onwards – of demonstrating these effects. This involved showing that town dwellers, in this case Londoners, experienced more of certain health complaints than others [12], with the underlying belief that smoke was the cause of this discrepancy:

. . . that her *Inhabitants* breathe nothing but an impure and thick Mist, accompanied with a fuliginous and filthy vapour, . . . corrupting the *Lungs*, and disordering the entire habit of their Bodies; so that *Catharrs*, *Phthisicks*, *Coughs* and *Consumptions* rage more in this one City than in the whole Earth besides [author's italics].

Medical opinion was not particularly supportive of this perspective. Although concurring with smoke being an infernal nuisance, the Royal College of Physicians (RCP) was displeased with Evelyn, regarding smoke as protective against infections. The king, however, responded to the tract by commanding Evelyn to prepare a bill to place before Parliament. The bill, which proposed that troublesome factories be removed many miles from London (to Greenwich and other areas down the river), was dropped, probably for economic and financial reasons [13].

More than a century later, in 1772, a new edition of *Fumifugium* was published, in which the editor laments the increase in smoke, invoking magistrates to take steps to check the evil, and urging investigation into production of a smokeless fuel. But nothing was done for 50 years, by which time the conversion of England from an agricultural to an industrial community had been substantial, as was the spread of smoke.

'As far as practicable': air pollution policy and public health in the nineteenth century

The industrialisation and urbanisation of the eighteenth century expanded almost exponentially during the nineteenth. These two elements compounded the smoke problem: growth of industry – itself creating increasing amounts of smoke – attracted people to the cities; and more people needed more coal for domestic heating.

Strong concern was evident from 1818, when a parliamentary committee was appointed to see what could be done in relation to smoke, in particular how far steam engines and furnaces could be constructed in a manner less prejudicial to public health and comfort [9]. The committee confidently noted the 'hope that the nuisance so universally and justly complained of may at least be considerably diminished, if not altogether removed' [14]. But these were wishes without solutions, and no further action was taken.

The middle of the nineteenth century witnessed the growing concerns with environmental and sanitary conditions outlined in the previous chapter. Chadwick's influential 1842 *Report on the Sanitary Conditions of the Labouring Population of Great Britain* has been mentioned in Chapter 2, and the following year a Royal Commission of *Inquiry into the State of Large Towns and Populous Districts* spelt out the terrible conditions in urban areas including poverty, overcrowding, congestion, crime and poor health. It was around this time that the conception of air in medicine really began to shift towards polluted air and its effects on health. As the century progressed, the shift became more and more apparent.

In the same year as the Royal Commission, a Select Committee examined the smoke problem and recommended introduction of legislation to control nuisances from furnaces and steam engines, and expressed the hope that black smoke, including that from private dwellings, might eventually be entirely prevented [9]. Just two years later another Select Committee reported that any bill to control smoke should be restricted to furnaces producing steam, and should not extend to fireplaces of common houses. As a result of this, in 1845, an updated Railways Clauses Consolidation Act contained a clause requiring railway locomotives to consume, 'as far as practicable', their own smoke. Two years after this a regulation applying to factory furnaces, with consumption requirements analogous to those applying to railways, was inserted into the Town Improvement Clauses Act, 1847, once again with the same qualifying phrase [13]. The City of London promoted a Sanitary Improvement Bill with the same clause in the late 1840s but the sixth version failed to become law due to industrial opposition.

It was around this period that municipalities began to grant authority to town councils to appoint a Medical Officer of Health (MOH), a physician with responsibility to examine the health of the local population, oversee provision of certain medical services and provide medical advice to the local authorities. For example, the Liverpool Sanitary Act of 1846, the first comprehensive sanitary measure passed in England, enabled the town council to appoint a MOH, and also a Borough Engineer and Inspector of Nuisances.

In 1848 the City of London appointed John Simon as MOH and, following the Metropolis Management Act, 1851, appointment of MOHs for various London districts became compulsory. In his 1850 annual report to the Commissioners of Sewers, Simon petitioned strongly for abatement of London's smoke. By 1856 there were 48 MOHs, and the same year the Metropolitan Association of Medical Officers of Health was formed. Other large municipalities made similar appointments – Leeds (1866), Manchester (1868), Birmingham (1872) and Newcastle (1873) – until the Public Health Act, 1875, made it mandatory for all districts to have a MOH

[15]. Two years earlier, with the number of MOHs outside London increasing, the Metropolitan Association of Medical Officers of Health expanded nationally to become the Society of Medical Officers of Health.

The MOH, working closely with the local population, saw at close hand the abject living and working conditions of the urban environment, and their effects on health. Furthermore, the work remit of MOHs placed them in a unique position to investigate and research the relationship locally between environmental conditions and health, and to act as a professional voice for improvement. With regard to smoke pollution the MOH was to embrace this dual role – as advocate and local researcher – and for decades to come a number of these figures spoke up passionately about the smoke evil, investigated links with health, and assisted pressure groups and campaigners. But, crucially, the lingering problem was what could actually be done about smoke? Whereas sewage and other sanitation issues could be more readily addressed, smoke was far harder, primarily because of its association with desired economic expansion.

One approach was to try to monitor more strictly the enforcement of legislation that already existed and create further powers if necessary. In 1853 the General Board of Health (GBH)[2] led an inquiry into inventions for the prevention of smoke, and recorded that a nine-hour observation of a cotton mill in Manchester showed smoke emission for 8 hours 52 minutes. The inquiry recommended the employment of police constables to make observations, and qualified officers to superintend them and to advise manufacturers [7].

In consequence of this report Lord Palmerston's Smoke Acts of 1853 and 1856, applying to London only, empowered the police to enforce provisions against smoke from furnaces used in steam raising, as well as smoke from other furnaces employed in factories, public baths and in furnaces used in the working of steam vessels on the Thames. These acts also contained the 'best practicable means' qualification, carefully nebulous wording that has remained embedded in environmental legislation.

Sanitary authorities in other parts of England and Wales were similarly empowered to take action in cases of smoke nuisances within the Sanitary Act, 1866. But this power was never exercised in London, where the police continued to try to enforce the Smoke Acts until the passing of the Public Health (London) Act of 1891 [16]. The Sanitary Act was repealed by the Public Health Act, 1875, within which legislation regarding smoke pollution was to be situated for the next 50 years.

The difficulties of enforcement of smoke legislation were multifold. First was the question of manpower. Both police officers and local authority officials were faced with time constraints and had to balance enforcement of the smoke laws with many other competing concerns. Next was the issue of skills, with the enforcers provided with training insufficient to monitor the enforcement of somewhat confusing legislation. Third was the matter of penalties, which were often too small to act as a significant deterrent – a theme that continued with subsequent legislation during the next century and is discussed in more depth later in this chapter. Fourth, it was impossible to actually measure levels of smoke in the air until air pollution monitoring networks were established around the First World War. And, finally, another recurring theme is that the legislation, dishearteningly for enforcers, focused mainly on tackling the problem after it had arisen, instead of specifying means of reducing smoke production *before* plumes were thrown out on a huge scale into the skies.

Despite commercial and industrial opposition to efforts to legislate against smoke pollution, there was support for reform from campaigners and, importantly, from the medical profession. Just a year after enactment of the second Smoke Act, the journal of the British Medical Association, the *British Medical Journal* (*BMJ*), reported a parliamentary question about whether the Government would actually enforce the 1856 act, which received the response that there had been 54 convictions in the previous six months with more expected [17].

Measuring the health effects of polluted air: statistics and research

Around this period was the emergence of journal discussion and research around the putative association of atmospheric pollution and poor health. Replacing earlier attempts at association, which had been largely anecdotal or theoretically framed, new statistical approaches were being developed, often using data on the number of deaths from certain diseases in each registration county: such data became more available subsequent to the 1836 Registration of Births and Deaths Act and the 1838 appointment of Donald Farr as Registrar-General.

For example, using mortality data along with local documents and population figures from the 1851 census, Bakewell examined mortality ratios from 'fever' in different districts. He suggested poverty (percentage of paupers) was most responsible for fever deaths, although probable amount of ventilation (average number of inhabitants in each house) was also important and, in a comment on causation, concluded that fever was due to 'food deficient in quantity and quality, especially as associated with filth and foul air' [18].

Bakewell's article of 1858 illustrates from an early date what has proven to be an ongoing difficulty for epidemiologists: how to prove a causal association between an environmental factor – such as atmospheric pollution – and health, when operating within a scientific (later biomedical) model of health. Connected to this point, the article also shows how air within Western medicine was starting to be conceptually fragmented in order to accommodate the new epidemiology: if it was not possible to measure air pollution directly, then a proxy indicator, in this case overcrowding as a measure of ventilation, could be used instead. In time, the possibility would arise of reducing air to smaller polluting components.

Although the 1850s were early days for epidemiology, the theoretical space was unfolding for medico-scientific investigation of the association between air and health, regardless of one's position on contagionism or miasmatism. At the end of that decade the Epidemiological Society held that some epidemic and endemic diseases have their origins (irrespective of specific poison) in atmospheric variations and, in their classification of disease, the third and last class included diseases originating in meteorological variations but were not transmissible [19]. Atmospheric and climatic conditions, in combination with polluted air, were believed to cause disease whether or not mediated by specific agents.

Urban areas were most heavily affected by the awful atmospheric conditions although, in comparison with previous centuries, town living had substantially improved in terms of health. The average life expectancy of a Londoner in the sixteenth century was 20, whereas in 1858 one could expect to live to the age of 37. But for the new medical scientists, and also the sanitary reformers, these figures were still unpalatable, life expectancy being much lower in the least healthy parts of towns. In a leading article of the Public Health section of the *BMJ*, the anonymous author refers to the large number of redundant deaths caused principally by 'foul air . . . , water and deficient light – the three spectres that are to be found in the cupboards of most poor men living in large towns' [20].

One of the difficulties facing MOHs was their position on the role of ventilation in combating the foul urban air. In some ways it was a losing battle. On the one hand it seemed that opening windows should be advocated, the air indoors being heavily polluted with domestic smoke and other emanations resulting from overcrowding. Generally, ventilation of the home as well as public buildings such as schoolrooms, was encouraged [21]. On the other hand, the outdoor air was often so bad that uncertainty existed over whether bringing it indoors was advisable.

Of course, home ventilation would be more desirable if ventilation of the towns could be improved. Writing in 1870 Oliver argues that the 'influence of the atmosphere on the general health and mortality of the inhabitants of towns, is beyond suspicion', and he promotes efforts made to purify the atmospheres of towns by encouraging street ventilation, and the application of legislative enactments to obtain the complete combustion of fuel, the opening of streets and alleys to allow ventilation of towns, and the formation of parks [22]. There was talk about the benefits of living by the sea, and the healthy influence of sea winds that followed river tracts into towns close to the coast. It had also been noted during a week of fog in 1873 that there were 700 more deaths in London than expected at that time of year [23].

Understandably, it was impossible to ventilate towns artificially when there was not the geographical predisposition to do so naturally; an option available to some town-dwellers in search of better health was to move to a purer atmosphere. Climatic treatments for respiratory complaints were gaining popularity and, for phthisical [tuberculous] patients, physicians were advising various climates outside towns including humid sea air, warm inland dry air, and the cold, dry atmosphere of the Alps. A short unauthored report in the *BMJ* in 1880 comments that if all these climates 'really have good effects on the health of tuberculous patients, we can only come to the conclusion that this happy influence is due only to the one factor which is common to them . . . , the purity of the atmosphere' [24].

The article also presented the results of German scientific investigations into how various climatic elements of 'pure atmosphere' might improve health, that had recently been published in the *Berliner Klinische Wochenschrift* and the *Revue des Sciences Médicales*. At an altitude of 1350 feet above sea level in Ottenstein, Saxony, investigators measured twice daily the 'respiratory affections' and 'rheumatic affections' of 106 patients with respiratory problems (35 chronic tuberculous, 14 chronic bronchial catarrh, 7 catarrh of the larynx) and 50 patients suffering from chronic rheumatism. Taking meteorological readings three times a day they found that for rheumatic patients the worst conditions were dampness and low mean temperature; for respiratory patients, aggravations in chronic phthisis; and in chronic catarrh they were coincident with cold days, considerable falls in the mean diurnal temperature, very humid atmosphere, and the predominance of cold northerly and damp west winds, as well as the richness of the air in ozone [24].

These investigations are particularly interesting as they illustrate how attempts were being made to correlate epidemiologically meteorological aspects of the atmosphere with measured morbidity, just as in the cities investigators would begin to turn towards correlating distinct elements of polluted air with ill health or death. But before correlation could be drawn, quantification of polluted air was needed, and in 1890 in the *BMJ* one of the first accounts appeared discussing the measurement of the amount of particles in air. The unauthored piece describes a communication made by John Aitken to the Royal Society of Edinburgh about an experiment in which air was tested for the amount of dust particles per cubic centimetre. It fails, however, to elaborate on how the air was tested or the particles counted.

The researchers found that the number of particles in the air – felt to affect the brilliance and transparency of the atmosphere – varied depending on wind direction. For example, in Cannes the number spread from 1500 when the wind was blowing from the mountain to 140 000 when blowing from the town. Importantly, particle numbers were shown to differ widely between urban and rural air [25], attributed largely to human activity:

Observations made in Scotland and elsewhere indicated how extraordinary was the pollution in the air due to human agency. In regions clear of human habitations, the number of particles fell as low as 200, while in and around villages the particles amounted to thousands, and in towns to hundreds of thousands.

Progress, however, in mitigating the smoke problem in the last quarter of the nine-teenth century was desperately slow. One reason was that smoke pollution was not viewed negatively by all. Some manufacturers and local residents felt comforted that 'the greater the smoke the greater the considered prosperity' [26]. In a similar vein, an article playfully entitled 'Cherishing the smoke demon' comments that although Londoners like to groan 'they cling on to smoke as a privilege of property' [27]. Meanwhile, as discussed in the previous chapter, some members of the medical profession maintained that impurity of the atmosphere corresponded with moral impurity, the immoral poor being blamed for neither caring sufficiently nor acting responsibly in relation to smoke pollution.

In the main, though, failure to improve the polluted atmosphere was due to lack of political and economic will, despite the emerging scientific evidence of illness being associated not only with outdoor air but also with air inside factories. Microscopic analysis had revealed small, rough jagged pieces of iron in the air of an iron factory, filaments of linen and cotton in shirt factory air, and Scottish mills were branded as 'human slaughter-houses' because of the spongy, spiky dust found in their air [28]. Uncertainty may have existed over whether these indoor dusts were inorganic or contained organic particles, but concern about the state of air within industries did lead to the Alkali, etc., Works Regulation Act, 1863.

Under this legislation, amended several times in subsequent years, a government-controlled inspectorate was created with the object of ensuring that all factory pro-cesses operated at the highest known level of efficiency. The 'Alkali Inspectors' acted as advisers and consultants to specific industries under their care, and imparted information immediately on any new methods for reducing effluent [29]. Decades later campaigners were still calling for a similar body to advise factories on the general problems of coal smoke production, separate to the restricted remit of the Alkali Inspectors.

As worrying as air quality inside factories was, those exposed represented (just a proportion of) the workforce, and a much bigger problem remained the pollution by coal smoke which affected urban communities on a huge scale. In fact, the extent of the problem seemed to be escalating. The legislative response, however, remained woefully inadequate.

The Public Health Act, 1875: landmarks and deficiencies

The Public Health Act, 1875, organised public health administration on a nation-wide basis. It divided the country into urban and rural sanitary districts subject to supervision of the Local Government Board (LGB), which had replaced the Privy Council in 1871. Existing local authorities fitted into the new pattern as far as possible, and borough councils became local health authorities, and, as mentioned earlier, it became mandatory for all districts to have a MOH.

Although the act may have been a landmark in the development of public health in England and Wales, it skirted carefully around the smoke problem. Provisions made within the new legislation to counter smoke were essentially two-fold. First, the act provided that any fireplace or furnace in trade premises must, so far as practicable, consume its own smoke. And, second, it enabled action to be taken – by an individual or local authority – against those responsible for factory chimneys emitting black smoke in sufficient quantity to be a nuisance [9].

A closer look at Section 91 of the act [30] shows the following listed as summarily punishable 'nuisances':

(a) factories, workshops, and workplaces 'not ventilated in such a manner as to render harmless *as far as practicable* [author's italics] any gases, vapours, dust or other impurities generated in the course of the work carried on therein that are a nuisance or injurious to health';

(b) fireplaces and furnaces which do not '*as far as practicable* [author's italics] consume the smoke arising from the combustible used therein' and are 'used for working engines by steam or in any mill, factory, dyehouse, brewery, bakehouse, or gaswork, or in any manufacturing or trade process whatsoever';

(c) 'any chimney (not being the chimney of a private dwelling house) sending forth black smoke in such a quantity as to be a nuisance.'

The main problem with this act was that every punishable offence had a broad caveat. Factories were required to ventilate their premises and use steam-raising equipment that consumed its own smoke, but only 'as far as practicable'. Writing in 1949 Jervis argued that 'best practicable means' became the loophole for routine legal defence, and did not require demonstration that much was actually being done to alleviate the smoke output. A growing range of new equipment was becoming available but it was expensive and, with uncertainty over how well it would work, was generally considered not worth the investment. Fines were cheaper [31].

For prosecutors, 'blackness' and 'nuisance' became the standards that needed to be proved. This was difficult since the smoke was often arguably grey and because divergent opinions existed in local authorities as to what constituted a nuisance. For instance, at the turn of the century Popplewell pointed out that in Bolton a fine could be imposed for three minutes in a half hour (or six in an hour) of dense, black smoke from a chimney, whereas in Oldham 12 minutes in an hour were allowed. He also felt that the fines were completely inadequate [5].

In addition, there were wide exemptions from the act. Section 91 contained the proviso that the courts must also have regard to the nature of manufacture or trade, and whether the fireplace or furnace had been carefully attended to or not; and Section 334 prevented the act interfering with the efficient working of mines, the smelting of ores or minerals, the calcining, puddling and rolling of iron and other metals, or the conversion of pig iron into wrought iron. To these processes more were added in 1926, along with any others that the Minister of Health could specify [29].

Perhaps most significantly, however, was that the act did not apply to private homes, known to be substantial contributors to smoke production. Although heating alternatives were limited at that time, there was soon to be an expansion in options, but the 1875 act set the tone for decades to come in not applying the legislation domestically. Moreover, no definition of 'private dwelling-house' was provided in the act, allowing further leeway for interpretation.

The act also did not apply to the capital, and similar provisions were not conferred until the Public Health (London) Act, 1891. An attempt had been made four years earlier to tackle London's smoke from private houses and other premises not included in the 1853 and 1856 Smoke Acts, when the Smoke Abatement (Metropolis) Bill was introduced to the House of Lords. But it had failed to become law [16].

By the late 1870s progress with air-pollution abatement was not as great as had been hoped. There were complications and confusion between existing acts and local laws, and legislative enthusiasm was waning.

The organisation of campaigning

Organised campaigning, however, was beginning on a much wider scale. In 1881, due to the extent of the smoke problem in London, the Kyrle Society and the National Health Society acted to form the Smoke Abatement Committee (SAC), a body aiming for the reduction of smoke from all sources but especially private dwelling-houses. The SAC organised what the historian Harold Platt has called a propitious event in urban environmental reform, the Smoke Abatement Exposition of 1881–2. Attracting over 115 000 visitors in London, and 31 000 in Manchester, the convention was a 'dazzling showcase for an international array of heating, lighting, cooking, and power hardware.' With a sliding scale of admission and popular lectures, the exhibition was also aimed at informing the working classes of the latest methods of domestic heating and ventilation [32].

Alongside organisation of important national events such as this, campaigning at the local level continued. The Leeds branch of the SAC prepared a petition in 1898, addressed to the President of the Local Government Board, pressing for an extension of the Alkali Act to include the government-led inspection of all factory chimneys. Although 'signed by a large number of influential and distinguished persons representing large industrial towns, trade councils, and architectural and medical societies', no action was taken by government [16].

Indeed the picture at the end of the century was rather bleak, in spite of the establishment in London of another pressure group, the Coal Smoke Abatement Society (CSAS) in 1899. To the frustration of many, alternative appliances for improved coal burning had been invented, and were duly being tested by organisations such as the

Smoke Abatement Institution and the Association for Testing Smoke Preventing Appliances. Reports of the testing were available but very little change in practice was actually occurring.

The MOH for Blackburn lamented that contrivances for admitting air had been 'invented and patented a thousand times' [33] and, ten years after the famous exposition, the mood is perhaps best summed up in a *BMJ* discussion of the 'old devil of factory smoke' [34]:

> The possibility of burning coal smokelessly and the principles which govern the operation of complete combustion are already widely known, not only among chemists, engineers . . . but among the manufacturers themselves, most of whom are perfectly well aware that smoke can be prevented, and how it can be accomplished . . . What is wanted is not so much additional tests . . . as considerably increased public interest in the matter, greater willingness on the part of the manufacturers to adopt the methods already proved efficient, greater independence on the part of smoke inspectors and local sanitary authorities, and, lastly, greater legislative powers to compel smoke producers to discontinue their present wasteful and injurious manner of obtaining heat.

NOTES

1. The poet Horace described discoloration of buildings in Rome, and Seneca wrote of the city's oppressive fumes.
2. The General Board of Health (GBH) was a central agency, established after the Public Health Act, 1848, with the role of guiding and aiding local authorities, and empowered to set up local Boards of Health. Chadwick was a member of the GBH, and John Simon was appointed Medical Officer to the Board. In 1858, medical functions of the GBH, including public health, were transferred to the Privy Council where they remained until 1871.

REFERENCES

1. B. H. Chen, C. J. Hong, M. R. Pandey and K. R. Smith, Indoor air pollution in developing countries. *World Health Stat. Q.*, **43**:3 (1990), 127–38.
2. K. R. Smith and S. Mehta, The burden of disease from indoor air pollution in developing countries: comparison of estimates. *Int. J. Hyg. Environ. Health*, **206**:4–5 (2003), 278–89.
3. World Health Organization Working Party, *The Right to Healthy Indoor Air* (Geneva: World Health Organization, 2000).
4. A. Marsh, *"Smoke": The problem of coal and the atmosphere* (London: Faber & Faber, 1947).
5. W. C. Popplewell, *The Prevention of Smoke Combined with the Economical Combustion of Fuel* (London: Scott, Greenwood and Co., 1901).
6. P. Brimblecombe, *The Big Smoke: A history of air pollution in London since medieval times* (London: Methuen, 1987).

7. A. Marsh, Air pollution legislation. In M. W. Thring, ed., *Air Pollution* (London: Butterworth's Scientific Publications, 1957), pp. 239–45.

8. *Ibid.*, p. 239.

9. P. C. G. Isaac, *The Clean Air Act, 1956* (Durham: University of Durham, Bulletin No. 8, 1956).

10. J. Evelyn, *Fumifugium, or the Inconvenience of Aer and the Smoake of London Dissipated* (London: Dorset Press, for the National Society of Clean Air, 1961). [Originally published in 1661.]

11. *Ibid.*, pp. 18–19.

12. *Ibid.*, p. 17.

13. C. Ainsworth Mitchell, Some medico-legal aspects of dust and its use as a means of identification in criminal investigation. *Medico-legal Criminolog. Rev.*, **IV** (1944), 195–9.

14. W. C. Popplewell, *The Prevention of Smoke Combined with the Economical Combustion of Fuel* (London: Scott, Greenwood and Co., 1901), p. xi.

15. G. Rosen, *A History of Public Health*, expanded edn (Baltimore: Johns Hopkins University Press, 1993).

16. Ministry of Health, *Committee on Smoke and Noxious Vapours Abatement: Final Report* (London: HMSO, 1921).

17. Anonymous, *BMJ*, 16 May (1857), 421.

18. R. H. Bakewell, A statistical enquiry into the causes of epidemics of scarlatina, measles, smallpox, and fever. *BMJ*, 11 December (1858), 1035.

19. B. G. Babington, Report of Societies [Epidemiological Society]: introductory address. *BMJ*, 20 November (1858), 973.

20. Anonymous, Public health. *BMJ*, 13 February (1858), 129.

21. F. H. Hartshorne, Ventilation of schoolrooms. *BMJ*, 30 June (1860), 512.

22. G. Oliver, The atmosphere of towns in its sanitary aspect. *BMJ*, 9 April (1870), 358–9.

23. P. Brimblecombe, *The Big Smoke: A history of air pollution in London since medieval times* (London: Methuen, 1987), p. 123.

24. Anonymous, The influence of climate on disease. *BMJ*, 10 January (1880), 56.

25. —, The presence and significance of atmospheric dust. *BMJ*, 8 February (1890), 313.

26. —, The smoke nuisance in manufacturing districts. *BMJ*, 19 April (1890), 911.

27. —, Cherishing the smoke demon. *BMJ*, 20 December (1890), 1442.

28. —, Ireland [Royal Irish Academy]. *BMJ*, 18 June (1870), 638.

29. R. A. Glen, *The Law Relating to Smoke and Noxious Fumes* (Manchester: National Smoke Abatement Society, 1934).

30. Public Health Act, 1875 (38 & 39 Vict. c. 55), s. 91.

31. J. J. Jervis, The combat with air pollution. In A. Massey, ed., *Modern Trends in Public Health* (London: Butterworth, 1949), pp. 140–80.

32. H. Platt, "Invisible demon": noxious vapour, popular science, and public health in Manchester during the age of industry. In E. M. Du Puis, ed., *Smoke and Mirrors* (New York: New York University Press, 2004), pp. 27–50.

33. S. Barswise, The abatement of the smoke nuisance. *BMJ*, 30 August (1890), 499–501.

34. Anonymous, The new attack upon factory smoke. *BMJ*, 7 June (1890), 1318.

4

Hot air but little action: air pollution policy and public health in Britain 1900–39

The first four decades of the twentieth century were an illuminating and formative period in the history of smoke pollution and policy. During that time the campaign against smoke pollution grew stronger, groups becoming more organised through affiliations and collaborations. But the face of the campaign needed to change. As the germ theory of disease took hold of scientific thought, medical interest in diseases lacking obvious microbiological aetiology waned. So the campaigners turned elsewhere, towards aesthetics, as well as taking a short side-step back to nature, in the guise of sunlight and greenery.

Visibility and sunshine

At the turn of the twentieth century it seemed little had changed. The engineer Popplewell captured the feeling of the times in his 1901 publication *The Prevention of Smoke Combined with the Economical Combustion of Fuel*. Here he complains that over the previous 80 years various acts concerning pollution have passed but 'in spite of all of this, the evil has been growing worse year by year, the law having to a great extent failed to effect any real improvement' [1].

Popplewell goes on to comment briefly on several of the themes that would dominate the debate on smoke pollution up until the Second World War. First, on the issue of whether industrial or domestic use is a larger contributor to the problem, he suggests the former dominates in manufacturing towns, the latter elsewhere, and apportions blame equally overall. Second, he puts forward the terrible damage done to buildings, citing Leeds Town Hall (which he held to be one of the finest buildings in the country), and speculates that the damage is effected by the visible portions of smoke, consisting of minute particles of carbon combined with a sticky, tarry matter. Following on from this, he paints a picture of blackened vegetation and lungs, the health of plants and humans alike suffering from smoke deposits. And last, he puts forward his concern about the loss of urban sunshine.

The formation of smoke, and accompanying loss of sunlight, was a developing area of interest for both meteorologists and anti-smoke campaigners. But whereas urban quality of life was the underlying motivation for campaigners, meteorological interest was generated by the desire to explain climatological phenomena scientifically – such as the formation of conditions leading to poor visibility – alongside growing concerns about the safety of flying.

Invented around 1850, the 'sunshine recorder'[1] was soon in general use and in 1880 the 'Weekly Weather Report' of the Meteorological Office included, for the first time, results from 16 stations around the country. By 1910 there were 131 stations, 66 of which had consistent readings for 30 years, and Frederick Brodie, a Fellow of the Royal Meteorological Society (RMS), presented results for these stations in the *Quarterly Journal of the Royal Meteorological Society* [2]. For many places in northern Britain, Brodie showed the mean duration of sunshine, over the 30 years between 1880 and 1910, to be less than one hour per day in December. For the month of December 1890 (the darkest month in the 30-year period), the sunshine record at a southern station, Westminster, was completely blank. The situation in large manufacturing towns was worst, especially in winter when rays were weak and domestic coal use high. The front cover of this book is a painting of the period by Monet, showing air pollution over the river Thames in London. Monet, and Whistler, were fascinated by the effects of smoke and fog on light.

Interestingly, however, Brodie points out that, over the three decades, the sunshine in London (stations at Kew, Westminster and Bunhill Row), although awful, had improved slightly. The mean percentage of possible light had increased from 26% to 31% over a five-year rolling period from 1881–5 to 1906–10, and, during the winter months, had increased from 7% to 14% at Westminster. The author suggests this change might be due to 'regulations affecting factories, to the improvements in the construction of domestic grates, and to the largely increased use of gas for cooking and heating purposes' [3].

Brodie's opinion that conditions were improving, in part due to legislation, but also to the activities of campaigners, were not held by all. Indeed Brodie himself had shown that the incidence of fogs in London (number of days with fog per year) had increased between 1870 and 1890 [4]. But meteorological interest in sunshine and visibility was gathering momentum [5]. In a 1902 edition of *Symons's Meteorological Magazine* J. E. Clark, another fellow of the RMS, describes his own scientific investigation of 'day darkness', a term being used increasingly in the City of London. From his desk at the Wool Exchange looking out over the Guildhall, Clark laboriously noted, between 1897 and 1902, the date and number of hours during which artificial light was used before a time reasonably near sunset. Correcting his observations to allow for absence at weekends, he calculated the number of quarter hours, and the number of days, that were dark each month.

Clark found the winter months worst, as expected, with the mean number of dark days over the five years highest in January (at 7), and the mean number of dark quarter hours highest in that same month (at 89). The figures for other winter months decreased with temporal distance from January, and Clark postulates that smoke plays a part in all three causes of darkness: low fog; high fog; and storms. The diurnal distribution of day darkness showed that artificial light was needed least in the middle of the day and most at 10 a.m. and 4 p.m., probably resulting from the excess of smoke due to lighting fires and lunch preparations. Clark calculated that, in a year, the amount of day darkness is equivalent to the same as the total office-hour time during winter between sunset and 5 p.m. In other words, half of the city's lighting expenses were due to day darkness [6].

So for the cities, which were most affected, there was a circular element to the problem. When it was cold, domestic coal fires were needed for warmth, which produced huge amounts of smoke. The smoke, as well as affecting health, kept out the sun's rays, creating darkness requiring the generation of electricity for artificial lighting. And the generation of electricity resulted in the further release of smoke into the atmosphere, exacerbating the problem. It does seem, however, that the incidence of fogs in London and other cities may have peaked in the 1890s. The slow decline thereafter, though hard to assess, may be attributed to lower pollution levels, changes in fuel type, tighter controls on industry and the work of campaigners [7].

Campaigns, exhibitions and committees

Within cities the problem was not that those with responsibility for monitoring smoke pollution were dismissive, but they were hampered by the limited legislation and so obliged to dwell repeatedly on investigation rather than action. Similar to previous efforts, the Chief Officer of the Public Control Department of the London County Council was asked in 1904 to prepare a report on how the smoke problem affected London, in which he again put forward the extent of the problem. Three years later the Public Health (Amendment) Act, 1907, addressed an element of the problem by enabling local authorities to make by-laws regarding the construction of chimney shafts for furnaces of steam engines and certain factories. But the legislation still only related to black smoke [8].

Owing to frustration with ineffectual legislation, officials began to work more closely with campaigners. A number of local authorities attended the 1909 Smoke Abatement Exhibition, held in Sheffield under the auspices of the Sheffield Federated Health Association, after which it was decided to create a National Smoke Abatement Committee (NSAC). Debate at the conference element of the exhibition suggested the most practicable and profitable solution would be the formation of a central Smoke Abatement Department under control of the Local Government

Board (LGB), containing expert inspectors who would advise local authorities and enforce the smoke abatement law – an arrangement similar to that of the Alkali Inspectors. But the LGB had declined an invitation to the meeting so the idea was shelved, and a further campaign organisation was formed, the Smoke Abatement League of Great Britain [9].

Around the country other towns were similarly keen to address the smoke problem, or at least show that they were trying to do so. The 1911 Smoke Abatement Exhibition in Manchester was reported in the *British Medical Journal* (*BMJ*) as designed to show the need for a cleaner atmosphere in Manchester and Salford, and to exhibit appliances that could reduce the amount of smoke emitted from chimneys. Smoke was calculated to cost Greater Manchester £1 000 000 a year – without taking into account damage to fabrics and its effect on health – with household fires responsible for 60% of the output. Promoting alternative fuel use, the piece goes on to suggest that 'if gas and electricity were sold at even a slight loss so as to induce people to use them instead of coal fires the gain in health and in other ways would be ample compensation' [10].

Just a year later the CSAS held another large display of possible solutions to the coal-burning problem – the International Smoke Abatement Exhibition – at the Agricultural Hall in Islington, London. At the end of this meeting a Committee for the Investigation of Atmospheric Pollution (CIAP) was formed which, though representing yet another 'expert group' set up to look at the problem, was notable for being a collaborative venture of the CSAS and the Meteorological Office,[2] signalling new relations in the quest for cleaner skies. *The Lancet* offered the committee its publishing and laboratory facilities, lending medical and scientific credibility.

The committee – which later became part of the Department for Scientific and Industrial Research and in 1956 was a committee of the Fuel Research Board – collaborated widely with local authorities and played an important part in establishing the daily recording of air pollution around the country; it eventually took control of air pollution monitoring. The degree of air pollution was measured as impurities suspended in the air – deposits which subside or are carried down by rain into gauges, and the photoelectric intensity of light reaching the ground (i.e. ultraviolet radiation). For example, early monitoring showed that for the years 1912–13 the average monthly fall of soot in London was 288 tons per square mile, and in 1920 an average of 17.6 tons of smoke per four-hour period were emitted from London chimneys in winter months [11].

Some authorities had independently started monitoring before the CIAP was set up, notably in Glasgow on the suggestion of the Glasgow Health Authorities, and in other large towns in Scotland [12]. While public health committees in individual authorities were usually responsible for carrying out the monitoring, the CIAP initiative stood out for its systematic coordination of monitoring of smoke output on a reasonably wide scale. This organised monitoring of air pollution indicators would prove of particular value because it allowed assessment of trends and changes over time.

The kind of collaboration that the CIAP programme generated had been fostered for some time by one of the committee's parent organisations, the CSAS. As well as being Honorary Treasurer of the new central committee H. A. Des Voeux was also an ardent campaigner for the CSAS, arguing in the organisation's 1912 publication *More Sunshine for London* [13] that darkness and dirt are close friends of disease, darkness favouring the growth of germs that sunshine kills. Incorporating psychological morbidity into the cycle of ill health, Des Voeux interestingly conveys a holistic dimension by adding that atmospheric gloom makes for mental gloom, and mental gloom for physical susceptibility to disease.

With sunshine and clean air the precursors of good health, it was the desire of the CSAS to 'arouse public opinion to the need of reducing materially the constant fouling of London's air with soot and sulphur from the crude combustion of coal' [14]. Des Voeux appeals for responsible action through householders realising that it is their duty to avoid polluting the community, through manufacturers adopting improved methods of combustion (substitution of a gas or electric motor for a steam engine, using gas-fired furnaces, improved training of stokers), and by individuals petitioning their local authority to punish offenders appropriately and also lobbying Parliament for more stringent legislation.

Yet despite the rousing words of such campaigners, the economic advantage of switching combustion machinery was not evident to industrialists or factory owners, and less polluting alternatives were beyond the means of most who lived in private homes in the cities. But alternatives were available, and it is worth noting that more than four decades before the landmark Clean Air Bill, Des Voeux stated categorically, albeit unsupported by figures, that gas fires were 'increasing in popularity to a very large extent' [13].

However, these were somewhat long-term goals. Change in public opinion would probably take some time, and dramatic change in coal use needed affordable smokeless fuels, which remained a distant hope. Although the surge in monitoring activity would in time enable more complex examination of smoke pollution, it did not help solve what was an immediate problem. The real issue was that incentives for change – both privately and in industry – were weak, and existing legislation was toothless in promoting or demanding change.

Such legal failings were emphasised by the Public Control Committee in a 1911 report recommending various alterations to the law [15], the main points repeating those made when the laws were passed:

- not restricting smoke nuisance to *black* smoke as in the 1875 Public Health Act (1891 for London);
- expanding the definition of a chimney to any opening through which smoke is emitted from a building where manufacturing goes on or where furnaces are used in operations carried out under statutory power, *including* any Government workshop or factory;
- extension of the power of sanitary authorities to take proceedings regarding nuisance *outside* of their area.

This last point stands out in reflecting growing recognition that pollution travels, and legislation needed expansion to allow local authorities to address smoke nuisance arising outside their immediate jurisdiction.

But the Government was not keen to change policy radically, a move that would be economically disadvantageous, in the short term at least. A Smoke Abatement Bill introduced by a private member to the House of Commons in 1913 did not proceed, and it was only when Lord Newton introduced a Smoke Abatement Bill to the House of Lords the following year that the Government felt obliged to take the matter more seriously.

Backed up by the corporations of many large boroughs expressing the need for tougher legislation directly to the LGB, Lord Newton's bill was withdrawn before a second reading upon assurance that the President of the Board would appoint, as soon as possible, a Departmental Committee with all interests represented that would examine carefully the existing law and its administration, and make proposals for consideration to Parliament [1]. That year, however, the country went to war and, although the President of the LGB, Sir Herbert Samuel, made appointments as promised, the work was suspended and the committee was not reconstituted until 1920. This time, however, it had the supportive and influential Lord Newton as Chairman.

The inter-war years

Lord Newton's Departmental Committee did eventually publish its investigative report in 1921 (Newton Report) following an interim version a year earlier. After over 50 meetings, interviewing 150 witnesses, and visits to many towns in England, Wales and Germany the inquiry blamed the prevalence of smoke pollution in this country on the indiscriminate and wasteful use of raw coal for all purposes, whether industrial or domestic, and on the lax administration of the law by the responsible authorities [9].

The report was especially severe on domestic pollution, calculating that about 2.5 million tons of soot from domestic fireplaces polluted the atmosphere annually (compared with 500 000 tons from industrial chimneys), a practice described as dirty, wasteful and unscientific. On the grounds of economy, loss of sunlight and damage to public health, the report recommended that this domestic practice should be restricted as much as possible.

But the committee fell disappointingly short of proposing significant legislative advance, advocating instead that the Government should encourage further research into domestic heating, that gas and electricity providers should be encouraged to cheapen supply, and that the practice of municipal authorities overcharging on gas and electricity in order to allocate profits to the relief of rates should be discontinued. The report weakly concluded that 'after full consideration, we do not consider it practicable at present to propose legislation dealing with smoke from private dwelling-houses' [16].

Instead, it was recommended that local authorities be empowered to make by-laws requiring provision of smokeless heating arrangements in new buildings other than private dwelling-houses and that 'the Central Housing Authority should decline to sanction any housing scheme submitted by a Local Authority or Public Utility Society unless provision is made in the plans for the adoption of smokeless methods' [17, 18].

On industrial smoke the report was similarly lacking teeth, advocating only that the Minister of Health's powers be extended (to act on a defaulting authority, and to fix emission standards from time to time), that duty to enforce be transferred to larger authorities, and that fines be increased. Paradoxically, the report suggested the 'best practicable means' imposed on manufacturers should also take account of cost, adding a further string to the defence bow.

According to the Chairman of the Smoke Abatement League of Great Britain, J. W. Graham, existing legislation in that area was, without any further caveats, already a dead letter. In an opening address to the 1924 Conference on Smoke Abatement in Manchester, he describes the legal requirement on stokers and manufacturers to avoid smoke production 'as far as is practicable' as unworkable and never used, as it leads to a capricious dispute about what is practicable, requiring expensive expert court testimony. As an example he cites how the London County Council unsuccessfully prosecuted the Lots Road Electric Station, and was obliged to pay costs of £300 because the magistrate declared the smoke was grey. Graham adds that the law also failed because fines were insufficient to act as a deterrent, and health committees and magistrates were often connected with the offenders socially or by family ties [19].

Developments in researching the health effects of air pollution

The CIAP had continued throughout the First World War with its plans to coordinate monitoring around the country and in 1916 produced its first report [20]. Summaries of the CIAP reports, which contained mainly data, were published in

The Lancet, lending medical gravitas to the largely fact-finding work. The wartime period delayed reports from observing stations but nevertheless by 1919 the fifth report was published [21]. Affiliation with the journal enabled some broad discussion, such as the 1922 seventh report's attention to visibility and colour [22]. It was not, however, until the final report that health issues were commented on.

At the end of this 1924 report the authors describe an attempt made to correlate atmospheric pollution with mortality in London by plotting the amount of suspended impurity in the air (daily impurity and six-day average) against number of deaths. They failed to find a definite relationship between pollution and mortality, although a relationship between death rate and minimum temperature 'appeared to be fairly definite'. The report concludes that a relationship between atmospheric impurity and deaths would likely only be seen over very long periods, but maintains that it is probable that 'the incidence of respiratory diseases and of fogs would show a definite relationship, although ... the low temperatures usually prevailing during foggy weather would be a complicating factor' [23].

These findings reiterate the already ongoing difficulties of demonstrating epidemiologically a causal relationship between an environmental factor, such as atmospheric pollution or a proxy for it, and a health outcome such as mortality. What investigators were finding was that common medical sense did not always tally with scientific analysis. It seemed obvious that smoke pollution was detrimental to health, yet trying to correlate the two over varying periods of time was proving largely unsuccessful. Even if an association was found, it remained difficult to be sure that another environmental factor, such as cold weather, was not at least partly responsible. Concluding that an observed association was causal was yet a further step.

An inability to drum up medical interest compounded the problem. Whereas atmospheric pollution occupied the minds and time of public health workers, MOHs, smoke inspectors, campaigners, meteorologists and the public, the medical profession was generally less taken. The germ theory of disease was widely accepted by this time, and interest in chronic diseases with environmental components was declining along with miasmatism. As the historian and health policy analyst Daniel Fox points out, in both the UK and the USA medical attention – along with health care and research funding – shifted heavily in the first half of the twentieth century away from chronic diseases and towards acute infectious diseases and the technological developments that accompanied this rapidly expanding area of knowledge [24].

An opening paper in a 1923 edition of the *BMJ* captures this change in orientation of the profession. While it had been accepted the previous century that insanitary conditions caused foul-smelling air which was responsible for disease, from the turn of the century attention was directed towards identifying the element(s) of foul air that might be responsible. But attempts to isolate micro-organisms in the foulest air, sewer and drain air, were proving fruitless. Extensive research from England, the USA and France in fact showed the opposite, that sewer air actually had a lower bacterial count than neighbouring atmospheric air, and there was no evidence of increased incidence of disease in sewer workers chronically exposed to the emanations. The author of the paper, W. E. Wynne, concludes that it is unreasonable to label sewer air as damaging, and argues that elaborate precautions using 'traps' to guard against reflux of sewer gas through drains into homes were misguided, wasteful and grounded in an imperfectly understood aetiology of epidemic and infectious diseases [25].

With medical interest diverted, and correlational analyses limited, researchers continued to consider the epidemiology of adverse events. This kind of descriptive epidemiology drew on bouts of particularly bad environmental conditions, notoriously episodes of smog, to demonstrate the corresponding short-term health effects. Although some of the same scientific problems still applied, the figures often spoke for themselves. While this approach will be looked at more in the next chapter, the opposite pole of the adverse event occurred when the 1921 coal strike presented a natural experiment in observing the effects of a dramatic reduction in smoke output.

Advocates of alternative, low-carbonisation fuels grasped the opportunity to point out that the death rate from respiratory disease in Glasgow during the 12-week strike period was half that of the year before [26]. Organised groups of women joined the voices heralding the healthy by-products of the strike, Lady Melville commenting in her pamphlet published by the Women's Printing Society *Choose Ye: Darkness or Light!* that 'probably for the first time since coal was generally used, we are enjoying, owing to the coal shortage, a pure atmosphere' [27]. The national newspapers announced improved urban sunshine [28], that 'fresh clean foliage refreshes the eye, and the scent of the lilac and May fills the air' [29], and declared the possibility of people seeing from the dome of St Paul's 'the vast town lying clearly before them, shining in the sun, with no pall of smoke overhanging its spires and myriad of streets' [30].

The 1926 Public Health (Smoke Abatement) Act

A combination of the Newton Report, the CIAP observational data, ceaseless campaigning, and a cross-sectoral deputation[3] to the Minister of Health, contributed collectively to the eventual legislative change [31]. Several smoke abatement bills were, in fact, discussed by Parliament between 1922 and 1924. Given the tone of the Newton Report, it was perhaps not surprising that the 1926 Public Health (Smoke Abatement) Act [32] was, in the minds of many, dreadfully inadequate [33].

Despite the broad support for significant change, the act was a watered-down hotchpotch of earlier ideas. Even the Newton Report recommendation, included in a 1924 Bill, that heating and cooking arrangements in new private dwelling houses be subject to any constructional by-laws, was rejected by the Minister of Health from the final act 'principally on the grounds that it would be unwise, in view of the national shortage of dwellings, to take any step, which might tend to hinder their erection or which might make them more costly' [34]. Somewhat defensively the Minister argued that many new buildings were being erected voluntarily with gas or electric cookers, and the number of open grates for the combustion of bituminous coal had been enormously reduced.

In other areas the act, which also applied to London, made some advances:

- it was no longer necessary to prove the smoke was black when emitted in such quantities as to be a nuisance;
- the definition of smoke was extended to include soot, ash, grit and gritty particles;
- it became the duty of local authorities 'to enforce the provisions of any act in force within their own district requiring fireplaces and furnaces to consume their own smoke' [35], and required them to inspect their districts and enforce the nuisance provisions generally, reinforced by granting the Minister of Health power to nominate a County Council to take over the task in an authority defaulting on its responsibility;
- the by-law-based powers of urban authorities were extended to include 'the provision in new buildings, other than private dwelling-houses, of such arrangements for heating and cooking as are calculated to prevent or reduce the emission of smoke' [36].

However, in significant areas the act was a disappointment. It failed to deal with domestic smoke at all, and Section 1(3) still allowed the same loophole for the industrial offender – in any proceedings against smoke other than black smoke – through the 'best practicable means' defence [37]:

… it shall be a defence for the person charged to show that he has used the *best practicable means* [author's italics] for preventing the nuisance, having regard to the cost and to local conditions and circumstances, and for the purposes of this subsection, the expression *best practicable means* [author's italics] has reference not only to the provision and efficient maintenance of adequate and proper plant for preventing the creation and emission of smoke, but also to the manner in which such plant is used.

Two additional, albeit less significant, points of the new act rankled with anti-smoke campaigners [38]. First, although fines were increased they still fell well short of their expectations. For example, the maximum penalty that could be imposed on persons against whom an abatement order was made went up from £5 under the 1875 Act to £50 – a ten-fold rise but too low to act as a deterrent, and far less than the cost of replacement combustion machinery. And, second, more industrial processes were *added* to those excluded from the law: reheating; annealing; hardening; forging; converting and carburising iron, as well as any other process specified by the Minister of Health in a provisional order [39].

So the new law was deemed to be largely ineffectual, an instrument to delay the introduction of policy that would create real change. In 1929 the National Smoke Abatement Society (NSAS) – a merging that year of the CSAS and the Smoke Abatement League – commented in the launch of its new journal that 66 local authorities now had by-laws under the 1926 act allowing industries a two- or three-minute period of black smoke emission per half hour, but most of these were from towns unimportant in terms of smoke nuisance [40]. Further, it was argued that these by-laws simply had the effect of standardising unwanted practice [41], and again set the tone for decades of focusing on a penal measure rather than preventing the conditions that gave rise to the smoke in the first place [42]. Despite powers given in the 1926 act, by 1931 no by-laws had been drawn up locally requiring provision in new buildings for heating and cooking arrangements to prevent or reduce the emission of smoke 'owing to the difficulty in formulating them satisfactorily' [43].

Animals, vegetables and filthy curtains

Those working to reduce smoke pollution during the 1930s drew largely on speculative or presumptive health effects, and increasingly the effects of smoke pollution on other aspects of human, animal and vegetable life. Linked to this, scientific research continued to break down smoke-polluted air into constituent components – essentially particles, gases and elements emitted from the imperfect combustion of raw bituminous coal – to which these adverse effects could putatively be attributed [44, 45].

With regard to human health, statistics (of the kind described earlier) were supported by pathological analyses. For instance, the lungs of city dwellers were histologically similar to those of coal miners,[4] and Taylor suggests smoke to be the cause. He cites, as evidence, that the 1928 infant mortality rate[5] for rural districts in England was far lower than that for county boroughs (55.5 vs 74.4), and that the infant mortality rate due to pneumonia and bronchitis showed a similar difference (8.8 vs 16.27). Taylor's view was supported by pathological evidence that grey, blotchy, sooty matter in the urban lung caused obstruction to air, in turn leading to less perfectly oxidised tissue more easily prey to lurking bacteria resulting in catarrh, bronchitis and emphysema, and ultimately cardiac failure triggered by acute circumstances [46]:

This affects more especially the very young and the old. After foggy weather the death rate from respiratory diseases rises enormously. It is highly probable that the high infant mortality in towns and especially the more congested parts is due to the filth in the atmosphere being too much for the delicate lungs of newly born children in conjunction with the effect of loss of solar radiation due to the smoke cloud hanging over towns.

Lack of the sun's ultraviolet light was also felt to cause rickets and Taylor, among others, believed the gloom to have serious psychological sequelae and a strongly negative impact on quality of life [47]. Comparisons with other countries were used to show the advantage of less smoke. In Germany, the relatively recent industrial advances had allowed the incorporation of newer, less polluting industrial processes, which was lauded. In New York, where the burning of soft coal was banned from 1905, the Health Commissioner claimed that the death rate from pulmonary TB had halved between 1905 and 1919 [48].

Animals were felt to suffer too. Dunn and Bloxham, for example, inferred that the deaths of more than 30 animals in a pasture adjoining a coke oven works in County Durham could be attributed solely to the lead from the smoke deposited on grass [11]. Polluted air, through its impact on grass, was also blamed for poor quality cows' milk, which in turn was postulated by the National Baby Week Council to affect child health. An experiment planting eight hollyhocks around Leeds showed that they flourished in direct proportion to distance from industrial areas [26]. While vegetation in cities was felt to be particularly affected by smoke, John Taylor (Assistant Medical Officer of Health for Manchester), writing in the *Smoke Abatement Handbook*, labelled both urban and rural vegetation as damaged by reduction of sunlight, solid deposits on foliage, acids and a soil surface of soot [49].

Alongside assessments of health effects, attempts were made to cost the damage done by smoke, especially to buildings and household contents such as curtains. In 1920 the Manchester Public Health Committee had calculated the extra cost incurred in cleaning materials (soap, starch, fuel) as $7^1/_2$d per week for lower rental households, higher costs in higher rental households, and with a conservative total estimate of £242 705 a year. A special further investigation into the washing of window curtains (washed on average every 11 weeks in Harrogate, North Yorkshire, but twice as often in Manchester) estimated 5 s a year per working-class house in Manchester, higher for larger houses, and a total extra cost for Manchester of washing curtains as £37 339 a year [47]. By 1929 soot deposits were calculated to cost the nation £80 million a year [26], and *The Times* reported in 1939 that the newly formed London Advisory Council for Smoke Abatement had calculated that, in addition to the centre of London receiving half the winter sunshine of nearby Kew, the annual 240 tons of soot deposited in each square mile cost the London County Council approximately £4 million a year [50].

But while costs of cleaning could always be viewed as unfortunate yet absorbable by-products of economic progress, there was some concern about the effects of poor visibility on flying. Work was being carried out by the Meteorological Office's Air Ministry on the conditions and mechanisms of bad visibility [51], and the Aviation Services Division of the same office showed that poor visibility increased markedly around aerodromes situated near large cities and industrial centres such as Croydon, Castle Bromwich (near Birmingham), Manchester and Alexandra Park where visibility was less than 2000 yards on one day out of two [52]. Despite the risks highlighted at a conference held by the National Smoke Abatement Society on Smoke and Aviation in 1935, air travel was essentially still low – that year there were only 40 departures and arrivals daily from London Airport at Croydon and about the same from Heston, an airport to the west of the city [53].

As the Second World War approached little had really changed in air pollution policy. A realisation of the rising contribution made by motor vehicles to pollution and ill health [41] had led to new legislation on the construction and equipment in motor vehicles [54], but the 1936 Public Health Act failed to do more than re-enact existing smoke legislation without any real material change [55]. Industries continued to pollute without much of a legal reason not to, and domestic coal use carried on unrestricted, change being determined purely by affordability and personal inclination. Some smokeless housing estates were beginning to appear [56], but many new buildings were still being built to accommodate coal. Change appeared far off, with Sir Hilton Young responding in Parliament (shortly before the 1936 act) to representatives from local authorities that he 'did not think it practicable to lay down restrictions for the elimination of domestic smoke as a condition of future housing subsidies' [57].

A quote from Herbert Clinch, Chief Smoke Inspector of Halifax County Borough, though taken from the 1923 *Smoke Inspector's Handbook* [58], encompasses nicely how the position on atmospheric pollution remained relatively unchanged as war broke out in Europe 16 years later:

The engineer is seeing in the smoke cloud a sign denoting inefficient boiler practice; architects and archaeologists are alike viewing with dismay the irreparable damage done to beautiful buildings... the city housewife, who wishes to keep open the windows for the sake of her children's health, finds herself unable to do it and keep her house clean at the same time ... With the Public Health worker the smoke abatement problem has long been a topic of urgent importance, and in more recent years he has been stimulated to action by the more precise knowledge which has been vouchsafed to him relating to the effect of sunshine, or the effects of its absence, upon the human organism.

NOTES

1. The sunshine recorder was originally a wooden bowl with a lens (first filled with water, then solid glass), the sun's rays charring the inside of the bowl, and charring was compared between bowls exposed for the same period. It was first used in 1855 and in 1879 Sir George Stokes replaced the bowl with a metal frame with grooves into which a strip of cardboard was inserted which received the rays and could be replaced daily [2].

2. The committee was chaired by Sir Napier Shaw, Director of the Meteorological Office.
3. Primarily concerned with the effects of smoke on child health the deputation consisted of representatives from the British Medical Association, Society of Medical Officers of Health, National Association for the Prevention of Infant Mortality, National League for Health, Maternity and Child Welfare, National Health Society, and the National Housing and Town Planning Association.
4. This was coined 'townsman's lung'.
5. The infant mortality rate, an important public health indicator, is the number of deaths under one year of age per thousand live births.

REFERENCES

1. W. C. Popplewell, *The Prevention of Smoke Combined with the Economical Combustion of Fuel* (London: Scott, Greenwood and Co., 1901), p. xi.
2. F. J. Brodie, The incidence of bright sunshine over the United Kingdom during the thirty years 1881–1910. *Quart. J. Royal Meteorological Soc.*, **XLII**:177 (1916), 23–35.
3. *Ibid.*, 35.
4. P. Brimblecombe, *The Big Smoke: A history of air pollution in London since medieval times* (London: Methuen, 1987), p. 111.
5. F. A. R. Russell, Further observations and conclusions in relation to atmospheric transparency. *Quart. J. Royal Meteorological Soc.*, **XXVIII**:121 (1902), 19–23.
6. J. E. Clark, Day darkness in the City. *Symons's Meteorological Mag.*, **XXXVI**:132 (1902), 194–6.
7. P. Brimblecombe, Urban air pollution. In P. Brimblecombe and R. L. Maynard, eds., *The Urban Atmosphere and Its Effects* (Air Pollution Reviews, Vol. 1) (London: Imperial College Press, 2001), pp. 1–20.
8. Public Health (Amendment) Act, 1907 (7 Edward II. c. 53).
9. Ministry of Health, *Committee on Smoke and Noxious Vapours Abatement: Final Report* (London: HMSO, 1921).
10. Anonymous, Smoke abatement exhibition. *BMJ*, 18 November (1911), 1378.
11. C. Ainsworth Mitchell, Some medico-legal aspects of dust and its use as a means of identification in criminal investigation. *Medico-legal Criminolog. Rev.* **XII**:IV (1944), 195–9.
12. Anonymous, The pure air problem. *BMJ*, 21 January (1911), 161.
13. H. A. Des Voeux and L. W. Chubb for the Coal Smoke Abatement Society, *More Sunshine for London* (London: Coal Smoke Abatement Society, McCorquodale & Co., 1912).
14. *Ibid.*, p. 8.
15. Anonymous, Smoke nuisance. *BMJ*, 28 January (1911), 222.
16. Ministry of Health, *Committee on Smoke and Noxious Vapours Abatement: Final Report* (London, HMSO, 1921), p. 17.
17. *Ibid.*, p. 18.
18. J. E. Stanier, *Report on Smoke Abatement* (Stoke-on-Trent: Gas Department of the City of Stoke-on-Trent, 1945).

19. J. W. Graham, *The Law Against Smoke* (opening session of the Conference on Smoke Abatement, Manchester, 1924) (Manchester: Smoke Abatement League of Great Britain, 1924).

20. Committee for the Investigation of Atmospheric Pollution, *Air Pollution (First Report)* (London: Office of the Committee for the Investigation of Atmospheric Pollution, 1916).

21. Advisory Committee for the Investigation of Atmospheric Pollution, *Air Pollution (Fifth Report)* (London: Office of the Advisory Committee for the Investigation of Atmospheric Pollution, 1919).

22. Committee for the Investigation of Atmospheric Pollution, *Air Pollution (Seventh Report)* (London: HMSO, 1922).

23. Committee for the Investigation of Atmospheric Pollution, *Air Pollution (Ninth Report)* (London: HMSO, 1924), p. 59.

24. D. Fox, *Power and Illness* (Berkeley: University of California Press, 1993).

25. F. E. Wynne, Discussion on the effect on health of sewer air and drain. *BMJ*, 27 October (1923), 760–3.

26. W. A. Bristow, *Smoke Pollution is Sapping the Vitality of the Race* (Leicester: The Reader Printing Co., 1929).

27. B. L. Melville, *Choose Ye: Darkness or light!* (London: Women's Printing Society, 1922).

28. *The National News* (26 June 1921). In B. L. Melville, *Choose Ye: Darkness or light!* (London: Women's Printing Society, 1922).

29. *The Times* (18 May 1921). In B. L. Melville, *Choose Ye: Darkness or light!* (London: Women's Printing Society, 1922).

30. *Evening News* (18 May 1921). In B. L. Melville, *Choose Ye: Darkness or light!* (London: Women's Printing Society, 1922).

31. Anonymous, [Untitled.] *BMJ*, 26 May (1923), 918.

32. Public Health (Smoke Abatement) Act, 1926 (16 & 17 Geo.V. c. 43).

33. E. Ashby and M. Anderson, *The Politics of Clean Air: Monographs on science, technology, and society* (Oxford: Oxford University Press, 1981).

34. Anonymous, The 1926 Act. *Clean Air* (National Smoke Abatement Society Journal), 2:1(Summer) (1930), 96.

35. Public Health (Smoke Abatement) Act, 1926 (16 & 17 Geo.V. c. 43), s. 92.

36. *Ibid.*, s. 5.

37. *Ibid.*, s. 1(3).

38. J. J. Jervis, The combat with air pollution. In A. Massey, ed., *Modern Trends in Public Health* (London: Butterworth, 1949), pp. 140–80.

39. Public Health (Smoke Abatement) Act, 1926 (16 & 17 Geo.V. c. 43), s.1(1)(e).

40. Anonymous, Introductory article (untitled). *Clean Air* (National Smoke Abatement Society Journal), 1:1(Autumn) (1929), 4.

41. R. A. Glen, *The Law Relating to Smoke and Noxious Fumes* (Manchester: National Smoke Abatement Society, 1934), p. 16.

42. A. Marsh, Progress review No. 12: atmospheric pollution. *J. Inst. Fuel*, September (1950), 1–4.

43. J. S. Taylor, Smoke and health. In National Smoke Abatement Society, *The Smoke Abatement Handbook* (Manchester: The Service Guild, for the National Smoke Abatement Society, 1931), p. 19.

44. Anonymous, Estimation of dust in air samples. *BMJ*, 6 April (1935), 709.

45. E. B. Hazelton, Carbon monoxide a predisposing cause of pulmonary tuberculosis. *BMJ*, 27 October (1923), 763–4.

46. Anonymous [editorial], A synopsis of a lecture on smoke and health. *J. Nat. Smoke Abatement Soc.* 1930; Autumn:119–120.

47. Manchester City Council (Air Pollution Advisory Board), *The Black Smoke Tax* (Manchester: Manchester City Council, 1920).

48. C. W. Saleby, More light on London (reprinted from *J. London Soc.*, July 1922, by the National Smoke Abatement Society) (London: McCorquodale & Co, 1922).

49. J. S. Taylor, Smoke and health. In *The Smoke Abatement Handbook* (Manchester: The Service Guild, for the National Smoke Abatement Society, 1931), pp. 13–21.

50. *The Times* (31 March 1939), Danger of London smoke pall: cost in health and money.

51. M. G. Bennett, Atmospheric pollution as affecting visibility. *J. Nat. Smoke Abatement Soc.* 1930; Autumn:123–127.

52. Entwistle F. Visibility as affecting aviation. *J. Nat. Smoke Abatement Soc.* 1930; Autumn:127–133.

53. W. Courtenay, Smoke abatement as it affects London. In National Smoke Abatement Society, *Conference on Smoke and Aviation* (30 May 1935) (Manchester: National Smoke Abatement Society, 1935).

54. Road Traffic Act, 1930 (20 & 21 Geo. V. c. 43).

55. Public Health Act, 1936 (26 Geo. V. and 1 Edward VIII. c. 49).

56. Anonymous, Smoke abatement. *BMJ*, 23 November (1935), 1030.

57. Parliamentary correspondent, Domestic smoke elimination. *BMJ*, 9 February (1935), 287.

58. H. G. Clinch, *The Smoke Inspector's Handbook or Economic Smoke Abatement* (London: H. K. Lewis and Co., 1923).

5

Disaster, reductionism and personal responsibility: air pollution policy and public health 1939–70

Production demands during the Second World War meant an inevitable increase in smoke pollution during that period. Industrial requirements continued after the war as a nation attempted to rebuild itself, bolstered by the community spirit that fostered regeneration needs over less immediate issues such as cleaner skies. But it did not take long for atmospheric concerns to regain momentum after the war.

The post-war period

In 1946 the awaited Report of the Fuel and Power Advisory Council, *Domestic Fuel Policy* (also known as the Simon Report), was published and brought back into view the need for attention to the smoke problem. Stressing that the moulding of public opinion was essential, the report recommended that any new town should be smokeless, and advocated by-laws necessitating prior approval (through certification) of new fuel-burning plants as well as their efficient maintenance. The report also encouraged the production of improved heating appliances (including domestic), as well as the introduction of minimum standards for such appliances, and advocated the need for adequate supplies of smokeless fuels at reasonable prices [1, 2].

Following this report, the second half of the 1940s saw the creation of new sets of powers – in the form of local legislation – to tackle pollution in a variety of ways. These powers are especially significant because of the historical tendency to emphasise the 1952 'Great Smog' in London as the catalyst of substantial change, the turning point in the battle against smoke pollution. Although that episode was undoubtedly important (as were the legal changes it engendered), what had been initiated beforehand is often overlooked.

Local authorities in industrial parts of the country began to press forward after the war by creating by-laws. Sometimes, prior legislation was needed to allow for new by-laws. For example, the City of London Act, 1946, extended provisions under the 1936 Public Health Act to make by-laws requiring that heating arrangements in new buildings, or substantial heating alterations in existing buildings, must include calculations that 'prevent or reduce to a minimum the emission of visible smoke' [3].

The by-laws often had several components and attempted to grapple with pollution from different angles. Stringent local regulations were sometimes placed on the production of industrial smoke. For example, an act passed in Birmingham in 1948 stipulated that no person 'shall install in any building . . . any furnace for steam raising or for any manufacturing or trade purpose unless such furnace is *so far as practicable* [author's italics] capable of being operated continuously without emitting smoke' [4]. Similar measures were enforced elsewhere, sometimes alongside new efforts to overcome the problem of (the distribution of) industrial smoke by stipulation of the minimum height of industrial chimneys.

This particular preventive initiative captures another enduring problem around measuring and understanding the health effects of air pollution – it is the nature of inhaled (ground-level) air that is most significant from the clinical epidemiological perspective, rather than either smoke output in a region or even atmospheric concentrations (neither of which may truly reflect what an individual actually breathes in). Nevertheless, it makes intuitive sense that getting the dirty smoke away from humans, higher into the skies, will *likely* be beneficial and, as an example, a by-law was drawn up in Dudley in 1947 which specified that every chimney erected in the borough 'be raised to such height measured from the level of the centre of the street nearest thereto as the Corporation shall reasonably require having regard to the use of such chimney the position of houses or other buildings near' [5]. Similar regulations about chimney height were enacted in other towns.

Of all the new by-laws the most significant class, however, contained those that involved the creation of areas in which smoke production was prohibited. Although often held up as the much-vaunted outcome of the 1952 smog and ensuing Clean Air Act, such areas were already being set up in a number of cities around the country. For example, the Manchester Corporation Act of 1946 stated that 'no smoke shall be emitted from any premises in the central area', demarcating that zone as 'bounded by the following highways or streets that is to say St Mary's Gate, Market Street, Piccadilly, Portland Street, Oxford Street, Peter Street and Deansgate.' The by-law continued that smoke included soot ash, grit and gritty particles, and the occupier of emitting premises would be liable to 'a penalty not exceeding ten pounds and to a daily penalty not exceeding five pounds' [6]. A similar smokeless zone was established in Crewe in 1949 [7].

The post-war period has additional pertinence as in 1948 the National Health Service (NHS) was established, with important implications for public health. These are looked at in the next chapter but it is worth noting here that public health was moved to local authorities, losing its role in the management of municipal hospitals (which essentially became State-run NHS hospitals), and being somewhat marginalised with the provision of community clinics.[1] Although involvement in environmental health matters continued, these changes heralded the start of what some have considered the decline of public health, or at least the diminishing of its status.

Still, in 1951, air pollution did not seem high on the Government's agenda. When the matter was raised in Parliament of whether enough was being done around smoke pollution, loss of sunshine and use of smokeless fuels, a respondent for the Ministry of Local Government and Planning commented that the Public Health Act, 1936, provided legal machinery, that smokeless zones were appearing and that there should be less domestic pollution on newly built housing estates [8]. But all was about to change.

The straw that broke the camel's back: the smog of 1952

There is no doubt that the awful smog[2] that occurred in London in the winter of 1952 was dramatically memorable for those that experienced it, and was important for its deleterious effects on human (and also animal) morbidity and mortality, as well as causing huge physical damage. But its central policy significance was probably more as the spark for change. After all, as discussed in earlier chapters, noteworthy legal changes were in process, smokeless fuels were already declining in price, and the frequency of fogs had been falling for some decades.

Furthermore, the smog of 1952 was certainly not a stand-alone event. There had been a number of documented severe fogs or smogs in the previous decades in British cities and overseas. For example, 60 people died in an episode in the Meuse Valley in Belgium in 1930, and at Donora, Pennsylvania, 20 people died and 7000 were ill during a fog in 1948 [9]. And there had been at least four previous notable London fogs or smogs. In 1873 a three-day episode in December was associated with a 1.4-fold increase in mortality in the week of the fog compared to deaths in the previous week (i.e. a mortality ratio of 1.4); a four-day episode in 1880 had a mortality ratio of 1.5; and a three-day episode in 1892 had a ratio of 1.3. Only four years before the infamous 1952 smog a record (in duration) six-day episode affected London and much of the country, and was associated with a mortality ratio of 1.3. In that episode, as usual, most extra deaths were attributable to bronchitis and pneumonia, but also some to cancer and myocardial degeneration. Most were in those aged over 45 but some increase was seen in pneumonia in infancy [10].

The 1952 episode caused by far the largest increase in deaths, a mortality ratio of 2.6, but the misery of the experience has perhaps also contributed to its longstanding prominence. Although many journal articles provide statistics of the effects of the smog, one of the most evocative descriptions of the happening itself was printed in the *Reader's Digest* in 1953, a condensed piece from an article that appeared in *La Croix du Paris* earlier that same year [11]. Although the abridged sections of the article below are still quite lengthy, their raw, human, scene-setting power is enduring:

On the afternoon of Thursday, December 4, 1952 there was nothing to indicate that this would be the Fog of the Century – that it would kill about 4,000 people, cause property damage of many thousands of pounds and bring the activities of the great metropolis almost to a halt.

By Friday morning a heavy, wet blanket had closed down. You could just see your own feet . . . As you groped along the pavement, blurred faces without bodies floated past you. Sounds were curiously muffled: motor-car horns, grinding brakes, the alarming cries of pedestrians trying to avoid the traffic and one another. This was a real "pea-souper", a "London particular" . . .

At London airport a few planes made instrument landings. One pilot, after landing, got lost trying to taxi to the terminal. After an hour a search party went out to look for him. But it too got lost. Soon all air traffic was suspended.

As the day went on, the fog changed colour. In the early morning it had been a dirty white. When a million chimneys began to pour coal smoke into the air it became light brown, dark brown, black. By afternoon all London was coughing.

On Saturday morning thousands of Londoners began to be frightened. They were those people, mostly over 50, who had a tendency to bronchitis or asthma. In a long black fog such people are in acute distress. Their lungs burn, their hearts labour, they gasp for breath. They feel as if they are choking to death – and sometimes they do.

By Saturday noon all the doctors were on the run. But there wasn't much to suggest – except to try to get to an oxygen tent. All hospitals were overworked . . .

Workers who couldn't get home slept in their offices or went to police stations and were put up overnight. Members of Parliament were issued blankets and bunked down in lounges of the House . . .

Police patrolled the docks in life jackets because people who couldn't see the ground walked into the water; a policeman at the Albert Docks pulled out eight. But too often the victims, though their cries were heard, couldn't be found . . .

On Sunday morning the fog was thicker than ever. At times visibility got down to 11 inches: literally you couldn't see your hand held out in front of your face . . .

It was cold that day. On the outskirts of town men and women, lost in the murk, sat down – and later were found dead of exposure . . .

Towards noon on Monday the fog lifted a little, then came down again. Then it rose a little more. Finally it was gone.

Londoners rubbed the soot out of their eyes and saw a city covered with dirt. Every piece of furniture had a slimy, black film. Curtains were so encrusted with soot that when they were cleaned they went to pieces. Blonde women became brunettes. It was weeks before the hairdressers and laundries and cleaners caught up with their work.

The 1952 great smog was clearly an awful event. Statistical analyses attempted to quantify the mortality, and sometimes the morbidity, attributable to the episode. The Chief Medical Statistician W. P. D. Logan reported in *The Lancet* that there were at least 4000 deaths during the two weeks following the start of the episode. Deaths assigned to bronchitis and pneumonia increased eight-fold and three-fold respectively in one week, and the overall mortality ratio of 2.6 showed most additional death to be in the older age groups, although some was in the very young: a ratio of 2.7 in those aged 75 and over; of 2.8 in those aged 45–64 and 65–74; of 1.8 in babies under four weeks old; and of 2.2 in infants aged four weeks to one year [12].

The speed of increase of deaths was astounding, rising markedly on the first day of the smog, 5 December, and peaking on 7 and 8 December. Most deaths, as in previous episodes, were assigned to bronchitis and pneumonia, but increases were also seen in lung cancer, coronary disease, myocardial degeneration and other respiratory diseases. The *BMJ* reported that the total deaths were more than double that of the two weeks before the smog, and more than treble that for the corresponding period in 1951 [13]. These deaths were felt to be additional, rather than simply brought forward [14]. Emergency hospital admissions for general acute cases rose in parallel over the period [15].

In addition to the damage to human health there were substantial material costs from accidents and filth, as well as economic costs such as production losses. There were also the costs of injury to livestock, other animals, plants and vegetation. Although many aspects of these were, by their nature, hard to quantify, some bizarre events occurred. For instance, prize cattle in the Smithfield Show fared particularly badly in the smog, with many deaths. It was speculated that these top animals did worse than their more mainstream counterparts because of the frequency of changing their bedding. Regular removal of straw also took away urine deposited in it, the ammonia content of which was apparently protective against the acidic smoke component of the smog. This conjecture was challenged at the witness seminar of a 2002 conference at the London School of Hygiene & Tropical Medicine that commemorated fifty years since the 1952 London smog. Pat Lawther stated at the seminar that research at the time suggested the animals actually died because they were too fat! [16]

What the 1952 smog appeared to be was something akin to the straw that broke the camel's back: people seemed to have finally had enough, and were genuinely scared by the event. This was accompanied by an apparent disbelief that nothing could really be done to attenuate the pollution problem, especially given the public's awareness of the growing availability of alternative fuels to coal; all this in the waning, but still present, post-war spirit of a fresh start and the opportunity to rebuild collectively.

The Clean Air Act, politics and policy

Pressed by campaigners and the vociferous public response to the smog, the government reacted in not atypical fashion by setting up a committee, under the

Figure 5.1   Thick smog in London. Traffic moves slowly, with lights aglow, as smog descends over the capital during daytime hours, 1953 (Bettmann/CORBIS).

chairmanship of Sir Hugh Beaver, to examine the national problem of smoke pollution – origins, causes, health and other impacts – and to make recommendations and provide policy options. The setting up of an expert committee to look at smoke pollution had been a regular political strategy for hundreds of years, so any sceptic could have been excused for being suspicious of another similar act.

Indeed Harold Macmillan (then Minister of Housing and Local Government) believed such problems to be both a matter outside the scope of governmental responsibility (domestic pollution was about personal behaviour) and an inevitable consequence of economic development, which was essentially desired and in the overall public interest. In a memorandum to the government in late 1953 [17] (Fig. 5.1) Macmillan wrote:

Today everybody expects the government to solve every problem. It is a symptom of the Welfare State . . . For some reason or another 'smog' has captured the imagination of the press and the people . . . Ridiculous as it appears at first sight I would suggest that we form a committee. We cannot do very much, but we can seem to be very busy – and that is half the battle nowadays.

The Beaver Committee, as it became known, was perhaps different to its predecessors. The chairman was a highly respected individual, which gave the committee crucial credibility and, along with the other committee members, worked tirelessly in gathering information. But, perhaps critically, Sir Hugh Beaver was also passionate about the subject, and seemed to hold convictions about the way forward, and the urgency of need for change, as he personally continued campaigning for years after publication of the report.

Set up in the summer of 1953 the Beaver Committee met frequently and took expert depositions. As well as looking at the historical, meteorological, epidemiological and economic evidence, the committee made trips overseas to compare and contrast the experiences of other countries. In 1954 the committee published its findings, which came to be known as the Beaver Report. Although not hugely innovative, the report was honest and tough on the problem of smoke pollution, documenting methodically the causes and consequences, and urging the need for intervention and change. The findings were broad, for instance including an estimation that the cleaning and depreciation of buildings (other than houses) cost about £20 million per annum, extra painting and decorating £30 million and corrosion of metals £25 million.

The committee made a number of recommendations (Table 5.1) [18]. The most significant of these were:

- prohibition of the emission of dark smoke (defined as darker than shade 2 on the Ringlemann Chart (i.e. 40% black) including, for the first time, domestic fireplaces;
- local authorities to be empowered to make orders to establish smokeless zones and smoke-control areas;
- obligatory arrangements to arrest dust and grit for certain industrial plant;
- the Alkali Inspectorate to be responsible for industrial premises with special technical difficulties around pollution control;
- financial assistance for conversion to smokeless fuel use;
- domestic heating appliances in new premises to be of approved types;
- the establishment of the Clean Air Council to coordinate and encourage research, and review progress made in implementing any legislation.

The groundswell of public opinion meant the findings of the report had to be taken seriously. This was reinforced when a private member's bill – introduced by Sir Gerald Nabarro, Conservative backbench MP for Worcestershire – was withdrawn after debate clarified that the Beaver Committee's findings had the general support of the House of Commons, and a comprehensive Government measure was promised [19]. The first draft of this measure appeared in July 1955 as the Clean Air Bill, and was debated. It was criticised by the Opposition and also Nabarro, with particular concern around the influence of the Federation of British Industry on the bill's drafting.

Table 5.1 The Beaver Report, Clean Air Bill and Clean Air Act: comparison of main areas

Beaver Report (1953)	Clean Air Bill (1955)	Clean Air Act (1956)
		Significant differences:
• *Emission of dark smoke*: prohibited from any chimney including domestic fireplaces (s.1).	• *Emission of dark smoke*: an offence from any chimney, except as permitted by regulations.	
• *Smokeless zones*: local authorities empowered to make orders, subject to confirmation by appropriate minister.	• *Smoke control areas*: local authorities enabled to establish 'smoke control areas', the new title for the old smokeless zones.	• 'Smoke control areas' not smokeless zones.
• *Industrial dust and grit*: obligatory plant for arresting dust and grit on certain installations (s.6); measurement of grit emission (s.7); grit emission should be minimised (s.5).	• *Industrial grit and dust*: new furnaces to be smokeless; local authorities to approve plans and specifications to indicate compliance with this provision.	• Domestic heating appliances in new premises were not obliged to be of approved type, and regulations around dark smoke did not apply domestically.
• *Special industrial premises*: where special technical difficulties, Alkali Inspectorate responsible for ensuring best practicable means for pollution prevention (s.17).	• *Special industrial premises*: the act will not apply to premises controlled under the Alkali, etc., Works Regulation Act, 1906; proceedings cannot be brought without consent of minister.	• The act would not apply to special premises under the Alkali, etc., Works Regulation Act, 1906, and proceedings could not be brought without ministerial consent.
• *Domestic heating appliances*: installed in new premises should be of approved types.	• *Grants*: for adaptation of fireplaces in private dwellings in planned smoke-control areas.	
• *Grants*: assistance by local authorities and Exchequer for updating appliances.	• *Smoke nuisances*: other than from a private dwelling or dark smoke from other building, may be dealt with under the Public Health Acts if a nuisance to local inhabitants.	
• *Railways and pit-heaps*: laws need updating; local authorities to enforce.	• *Building by-laws*: in future may be required with regard to heating and cooking to prevent, as far as practicable, the emission of smoke.	
• *Annual report*: on smoke abatement required by local authorities to appropriate minister.	• *Research and publicity*: gives local authorities powers to promote or assist research, and engage in education and publicity.	
• *Penalties*: for smoke offences increased.		
• *Clean Air Council*: set up, chaired by Minister of Housing and Local Government, or in Scotland by the Secretary of State.		

The bill had an extended parliamentary stay, with numerous drafts and alterations along its chequered path to legislation [20]. By the time the bill was enacted as the Clean Air Act, some important elements had been either removed or significantly tailored (see Table 5.1) [21]. The main differences were:

- that smokeless zones were abandoned in favour of the less-stringent concept of smoke control areas in which the emission of chimney smoke would constitute an offence, and smokeless fuels must be burnt unless fuels capable of emitting smoke could be burnt smokelessly;
- that there was no provision in the act requiring domestic heating appliances in new buildings to be of approved types, and regulations around dark smoke did not apply domestically;
- that the act would not apply to industrial premises covered by the Alkali Inspectorate;
- omission of the Beaver Committee's recommendation that the Government Loan Scheme for fuel-saving projects be extended to include projects specifically directed to secure the reduction of air pollution;
- omission of the Beaver Committee's recommendation that purchase tax on gas and electric heaters be abolished;
- that in planned smoke control areas any owner or occupier incurring expenditure on the adaptation or replacement of a fireplace or chimney would be entitled to repayment of 70% of the cost by the local authority (who could themselves recover 40% of the total cost from the Exchequer and could also repay the whole or part of the remainder). The owner was therefore left with paying up to a maximum of 30% of the cost.

The Clean Air Acts, 1956 and 1968

Finally passed in 1956, the Clean Air Act still promised much, but it would take time to deliver [22]. As has been outlined above the main contentions were that, although offering improvement in industrial smoke output, the act still left domestic pollution relatively unchecked, and in the hands of local authorities. However, given that the domestic chimney was felt to contribute almost half of atmospheric pollution, many lamented this abrogation of responsibility, and there was no specific attention paid to increasing availability of smokeless fuels at a reasonable price.

Smoke control areas were a very watered-down version of smokeless zones. It was likely that the impact of legislation that was clearly prohibitive, would be significantly less than the impact of legislation that promised to do the best it could. Only chimney emissions were under the new remit, and the burning of garden fires and industrial waste in the open were both allowed [23]. Penalties would be fines (maximum of £10 a day), incurred at the discretion of the local authority. A memorandum on the subject [24] clarified the matter:

The effect of the Smoke Control Order, broadly speaking is to prohibit entirely the emission of smoke from chimneys in a certain area; but the provisions of the Act are flexible and allow for adaptation to local circumstances. Smoke control areas may be completely smokeless areas, like the smokeless zones which some local authorities have already established under local Acts, in which case all buildings are controlled; or they may be areas, perhaps larger in extent, in which certain classes of buildings only are exempt, so that the area as a whole will not be entirely smokeless.

The act was, perhaps, a reflection of the huge swing that would gain pace in the 1960s towards personal responsibility, for both lifestyle choices (such as around sexual freedom with development of the contraceptive pill) and for matters related to health. In the smoke pollution domain, the onus moved squarely onto individuals to make the domestic changes recommended in the Clean Air Act, although the changes became easier with the support of local authority grants and the falling prices of smokeless fuels. This will be returned to in the next section.

Progress was slow but steady through the late 1950s as specific elements of the Clean Air Act began to be implemented and the general ramifications filtered through. A questionnaire review of local authorities undertaken by the National Society for Clean Air[3] in 1960 showed that the prevention of smoke from industrial processes was improving with better fuel efficiency and new techniques, that industrial contraventions were being reported, and that many plans for chimney heights had been submitted. In that year there were 157 smoke control areas in operation, with another 587 orders confirmed or submitted, compared with the 44 smokeless zones that existed in 1956 [25].

The Government issued circulars in 1962 asking local authorities to speed up their smoke-control programmes, and reiterated that arrangements had been put in place to substitute smokeless fuel or cash payments for coal, in order to alleviate the monetary concerns of those receiving coal at concessionary rates from the National Coal Board [26, 27]. That same year a White Paper on cleaner air, *Smoke Control*, was produced. Some local authorities, however, still did not envisage completing their smoke control programmes until the mid-1970s (Fig. 5.2).

The Clean Air Act, 1968, extended the 1956 act, making it an offence to emit dark smoke from industrial or trade premises (the emission of dark smoke from chimneys was covered in the earlier act, although limited periods were allowed), and amended the requirement for certain kinds of industrial furnace to be fitted with grit and dust arrestment plant approved by the local authority [28]. By 1970 London had not had a smog in eight years, and many suspected would not have another one.

Figure 5.2   A policeman from the City of London Police Force wears a face-mask to protect himself against the city smog in 1962 (Hulton-Deutsch Collection/CORBIS).

Economics, smoke and smokeless fuels

Although the 1952 smog might have galvanised policy change, the changes it engendered were effective because they took place in the context of already levelling coal consumption and smoke production, and the increasing availability of alternative fuels. It was a blip on the improvement curve, really propelled forward in the 1960s by issues of economics and personal behaviour.

A look at the production and use of coal is illuminating (Table 5.2). Coal production rose dramatically with increasing population size and changing industrial patterns from around 1700 onwards. Coal output (coal produced for national use and for export) rose from 10–15 million tons a year in England in 1800 to 287 million tons a year in 1913, before falling somewhat with the advent of the First World War. By the outbreak of the Second World War 180 million tons a year were being consumed nationally, 28% of which were for domestic (home) use. In 1953, after the 1952 smog but prior to the Clean Air Act, total coal consumption had levelled off to just over 200 million tons a year, the domestic contribution of which had fallen to around 36 million tons a year (18%). It was not really until the

Table 5.2 Trends in coal use, smoke production and pollutants in England

Year	Coal output [O] / *or* consumption [C]			Smoke production		
	Total (million tons/annum)	Domestic million tons/annum (% of total)	Industry million tons/annum (% of total)	Total (million tons/annum)	Domestic million tons/annum (% of total)	Industry million tons/annum (% of total)
1700	3 [O]					
1800	10–15 [O]					
1900	220 [O]					
	160–70 [C]					
1913	287 [O]					
1925	230 [O]					
1936–8	180 [C]	51 (28%)	129 (72%)[a]	2.3	1.2 (52%)	1.1(48%)[a]
1948	190 [C]	37 (19%)	153 (81%)[a]	2.0	0.9 (45%)	1.1 (55%)[a]
1952	224 [C]	36.8 (16%)	187.2 (84%)[a]	2.3	1.3 (57%)	1.0 (43%)[a]
1953	205 [C]	36.8 (18%)	168.2 (82%)[a]	2.1	0.9 (43%)	1.2 (57%)[a]
1958	200 [C]	36.0 (18%)	164 (82%)[a]	1.7	1.1(65%)	0.6 (35%)[a]
1959	221 [C]	33.6 (15%)	187.4 (85%)[a]	1.9	1.2(63%)	0.7 (37%)[a]
1964	184.7 [C]	27.9 (15%)	156.8 (85%)[a]			
1967	161.7 [C]	23.0 (14%)	138.7 (86%)[a]	0.87	0.75 (86%)	0.12 (14%)[a]

[a] As well as industrial works this includes railways, collieries, gas works, coke ovens and electric power stations. For the sources for this table see references 52–59.

mid-1960s that total consumption began to fall more substantially, with the domestic contribution falling proportionally at slightly greater speed than the industrial component.

Of particular importance to atmospheric pollution, however, is the smoke produced from the coal consumed. Here the proportional contribution from domestic smoke is key. As Table 5.2 shows, although domestic coal consumption through the 1940s, 1950s and 1960s made up only 15–20% of the total coal consumed, this domestic use consistently contributed 45–65% of total atmospheric smoke pollution. By the late 1960s this proportion had risen, despite a fall in total smoke production. The enduring apparent mismatch existed because of the type and quality of the coal used (for instance, whether it was cleaned first), and the nature of the coal-burning apparatus – the domestic fire.

So domestic fires were highly significant to the Beaver Committee's estimated £250 million of damage caused by smoke pollution – this was felt to be a conservative approximation [29]. But the sources and availability of home energy had been changing. Domestic oil and gas use increased 50% between 1938 and 1956, with the rise in solid smokeless fuels only slightly less. Domestic electricity use in that period,

8. In 1927 the Advisory Committee on Atmospheric Pollution transferred from the Meteorological Office to the Department for Scientific and Industrial Research (DSIR). In 1945, the Atmospheric Pollution Research Committee was reconstituted as a Committee of the Fuel Research Board of the DSIR. Work on atmospheric pollution at the DSIR, including the monitoring networks, involved close cooperation with the Meteorological Office, Ministry of Housing and Local Government, Ministry of Fuel and Power, Ministry of Health, Medical Research Council and, of course, the local authorities.

REFERENCES

1. A. Wade, Atmospheric pollution: the present position and the outlook for the future. *Sanitarian*, **56**:6 (1948), 157–9.
2. F. J. Redstone, *Atmospheric Pollution*, Sanitary Inspectors Association Annual Conference, Blackpool, 1948, Paper No.6, 24.
3. City of London (various Powers) Act, 1946 (9 & 10 Geo. X. c. 29) s. 14 (I).
4. Birmingham Corporation Act, 1948 (11 & 12 Geo. VI. c. 39), Part IV, s. 46 (I).
5. Dudley Corporation Act, 1947 (10 & 11 Geo. 6. c. 27), Part IV, s. 40 (I).
6. Manchester Corporation Act, 1946 (9 & 10 Geo. VI. c. 38), Part V, s. 35 (I).
7. Crewe Corporation Act, 1949 (12 & 13 Geo. VI. c41), Part VI, s. 50.
8. Medical Notes in Parliament, Smoke pollution, *BMJ*, 10 March, 534.
9. Political and Economic Planning (PEP), *The Menace of Air Pollution* (London: Chiswick Press (for PEP), 1954, Vol. XX, 369, 189–215).
10. W. P. D. Logan, Fog and mortality. *Lancet*, 8 January (1949), 78–9.
11. E. Muller, The great London fog. *La Croix de Paris*, 6 March (1953) (condensed version in *Reader's Digest*, June (1953), 125–30).
12. W. P. D. Logan, Mortality in the London fog incident. *Lancet*, 14 February (1953), 336–8.
13. Anonymous, Deaths in the fog. *BMJ*, 3 January (1953), 50.
14. E. T. Wilkins, Air pollution and the London fog of December, 1952. *Sanitarian*, February (1954), 224–34.
15. G. F. Abercrombie, December fog in London and the Emergency Bed Service. *Lancet*, 31 January (1953), 234–5.
16. The big smoke: fifty years after the 1952 London smog – a commemorative conference. Held at the London School of Hygiene & Tropical Medicine, December 2002. www.lshtm.ac.uk/history/bigsmoke.html (accessed 17 May 2005).
17. M. Hamer, Ministers opposed action on smoke. *New Scientist*, 5 Jan: 1391 (1984), 3.
18. P. C. G. Isaac, *The Clean Air Act, 1956* (Durham: University of Durham, Bulletin No. 8, 1956).
19. Anonymous, The Clean Air Bill. *Coke & Gas*, September (1955), 1–10.
20. Standing Committee B (Parliamentary Debates), *Clean Air Bill: thirteenth sitting* (London: HMSO, 1956).
21. J. A. Scott, *The Clean Air Bill and the Local Authority*, London Sessional Meeting of the Clean Air Bill, 17 October 1955 (London: Royal Society for the Promotion of Health, 1955).

22. H. Beaver, *Clean Air: The next chapter*, The seventh Des Voeux memorial lecture of the National Smoke Abatement Society Annual Conference, 1956 (London: National Smoke Abatement Society, 1956).

23. A. Gilpin, *Clean Air: An economic, social and administrative problem*, Paper no. 6 presented at the Annual Conference of the Association of Public Health Inspectors, Eastbourne, 1957.

24. Ministry of Housing and Local Government, *Clean Air Act, 1956: Memorandum on smoke control areas* (London: HMSO, 1956), p. 3.

25. A. Marsh, Progress review No. 48: air pollution. *J. Inst. Fuel*, December (1960), 609–15.

26. Ministry of Housing and Local Government, Smoke control areas, *Clean Air Act, 1956*, Circular No. 3/62 (London: HMSO, 1962).

27. Ministry of Housing and Local Government, Smoke control in the black areas, *Clean Air Act, 1956*, Circular No. 4/62 (London: HMSO, 1962).

28. Ministry of Housing and Local Government/Welsh Office, *Clean Air Act, 1968*, Joint circulars 72/69 and 71/69 (London/Cardiff: HMSO, 1969).

29. R. S. Scorer, The cost of air pollution from different types of source. *J. Inst. Fuel*, March (1957), 110–15.

30. A. Parker, Air pollution. *Chem. Ind.*, 25 June (1966), 1129–31.

31. P. G. Sharp, *Towards Cleaner Air – A survey of air pollution* (Brighton: National Society for Clean Air, 1968).

32. Committee on Solid Smokeless Fuels, *Report of the Committee on Solid Smokeless Fuels* (London: HMSO, 1960).

33. Ministry of Housing and Local Government, *Clean Air Act, 1956*, Circular No. 69/63 (London: HMSO, 1963).

34. Ministry of Power, *Domestic Fuel Supplies and the Clean Air Policy* (London: HMSO, 1963).

35. Ministry of Housing and Local Government, *Clean Air Act, 1956*, Circular No. 13/65 (London: HMSO, 1965).

36. The Consumer Advisory Council, *Shopper's Guide 26: Smokeless Fuels* (London: The Consumer Advisory Council, 1962).

37. R. C. Avery, Economic aspects of air pollution and its control. *Clean Air*, 7:27 (1977), 26–30.

38. Anonymous, Air pollution and lung cancer. *BMJ*, 1 November (1952), 982–3.

39. P. Stocks and J. M. Campbell, Lung cancer death rates among non-smokers and pipe and cigarette smokers: an evaluation in relation to air pollution by benzyprene and other substances. *BMJ*, 15 October (1955), 923–9.

40. A. S. Fairbain and D. D. Reid, Air pollution and other local factors in respiratory disease. *Br. J. Prev. Soc. Med.*, **12** (1959), 94–103.

41. A. S. Fairbain and D. D. Reid, Air pollution and other local factors in respiratory disease. *J. Epid. Comm. Health*, **51** (1997), 216–22.

42. Expert Committee on Environmental Sanitation (Fifth Report), *Air Pollution* (Geneva: WHO, 1958).

43. E. C. Halliday, *A Historical Review of Atmospheric Pollution* (Geneva: WHO, Monograph Series No. 46, 1961).

Page 99 References

44. Report of a WHO Expert Committee, *Environmental Change and Resulting Impacts on Health*, WHO Technical Report Series No. 292 (Geneva: WHO, 1964).

45. P. J. Lawther, A. E. Martin and E. T. Wilkins, *Epidemiology of Air Pollution* (Geneva: WHO, 1962).

46. P. J. Lawther, Air pollution and the public health. *J. Royal Soc. Arts*, September (1965), 744–52.

47. Committee of the Royal College of Physicians of London on Smoking and Atmospheric Pollution, *Air Pollution and Health* (London: Pitman, 1970).

48. Anonymous, Progress with air pollution. *BMJ*, 2 September (1967), 570.

49. [Untitled], *BMJ*, 2 June (1962), 1547.

50. A. C. Saword, Environmental public health services. *Med. Press*, 26 June (1957), 581–6.

51. British Medical Association, *Clean Air* (London: BMA, 1965).

52. A. Parker, Cities without smoke. *J. Roy. Soc. Arts*, December (1950), 1–17.

53. —, Smoke abatement. *Sanitary Inspectors Association Annual Conference Brochure*, Paper No. 6:2 (1950).

54. Political and Economic Planning (PEP), *The Menace of Air Pollution*, Volume XX (London: Chiswick Press, for PEP, 1954), pp. 189–215.

55. A. C. Sawford, Clean air. *Med. Press*, 26 June (1957), 581–6.

56. A. Marsh, Air pollution (Progress Review No. 48). *J. Inst. Fuel*, December, 609–15.

57. S. R. Craxford, Air pollution – past, present and future. *Inst. Petroleum Rev.*, **15**: 173 (1961), 134.

58. A. Parker, Air pollution. *Chem. Ind.*, 25 June (1966), 1129–31.

59. P. G. Sharp, *Towards Cleaner Air: A survey of air pollution* (Brighton: National Society for Clean Air, 1968).

Conclusion to Part II

The conceptual relationship between air and health now looks rather different to that presented in Part I. The harmonious orientation of mankind and nature, with its holistic and sometimes spiritual dimension, disappeared with the earlier civilisations, to be replaced with a more practical notion of air as part of the natural environment. A synergy emerged in the Enlightenment and through the Industrial Revolution of polluted air as the environmental medium of infectious diseases according to miasmatic theory, and 'seeing' the more direct ill-health effects of air as those visible in the atmosphere as smoke pollution.

Advances in medical science and acceptance of the germ theory of disease led to a de-emphasis on the place of the environment in relation to health. Developments in public health in Britain displaced attention further from environmental health, and from improving the natural environment as a means to improving the public's health. Smoke pollution came to symbolise concerns about the atmosphere, yet reduction of smoke pollution was not politically palatable while industrial development was paramount. Eventually, falling prices of alternative fuels to coal, market competition, and a public tired of choking smogs and willing to take more responsibility, combined to turn the tide.

In the mid-1970s the epidemiological understanding of the effects of air pollution on human health had progressed, but to a limited degree and in a certain kind of direction. Infants, children, the elderly, certain occupational groups and those with pre-existing respiratory or cardiovascular disease were most susceptible to air pollutants, particularly in acute pollution episodes. But this had been observed for a long time.

The extent to which polluted air causes long-term ill health, however, remained more difficult to quantify and more poorly understood. But this was not for lack of epidemiological studies, and a reductionistic understanding was emerging for both the acute and chronic effects of pollution, one of the associations between smaller component pollutants in air and various measures of ill-health. This understanding was sufficient, in fact, for the development of standards relating to concentrations of air pollutants categorised by their risk level to health.

Air Pollution, Epidemiology and Public Health: Theoretical and Philosophical Considerations

Overview of Part III

Parts I and II of this book charted the changing conceptions of air and health from ancient civilisations to modern society. Part I illustrated how in Greek medicine in particular – but also in other early medical systems – air fitted into a holistic framework of health and illness which contained mankind and the environment, and which was often imbued with spiritual elements. By the end of the Enlightenment, however, spiritual dimensions had been largely replaced by scientific rationalism in Western medicine: increasingly dominant theories of miasma placed air within the new science as the conveyor or creator of disease, but this concept was framed by a larger debate about the environment, evolution and morality.

Part II picked up the story in the middle of the nineteenth century, and examined how air in medicine became equated with polluted air, especially smoke, and its effects on human health. Over the following 100 years, despite evidence provided by descriptive epidemiology, both public health and epidemiology had little impact on efforts to ameliorate air pollution and its health effects. Economic and political considerations characterised a century of policy procrastination punctuated by only one significant legislative event, in the face of concerted efforts made by increasingly organised campaigners for cleaner skies.

Part III continues chronologically, and covers a period of three decades up to the present day during which subtle but highly important shifts have occurred in the way air is conceptualised in epidemiology and public health. To explore these shifts a case-study approach is employed, involving a quantitative risk assessment (QRA) of the impact of air pollution on health in an area of north London. The method of QRA has developed against a backdrop of a renaissance of interest since the 1970s in the natural environment and its impact on human health, underscored by movement towards recognising the importance of human interaction with the natural, physical environment.

Air – still in the guise of polluted air – has fitted into this resurgence. The case study is presented in Chapter 7 and illustrates how methodological advances have seen air further reduced, or re-conceptualised, to components of polluted air around which the scientific search for associations with ill health has focused.

In Chapters 8 and 9 some of the philosophical and ethical issues that both constrain and dictate current epidemiological thinking and public health practice are then explored: problems with causality; lack of coherent epidemiological theory; the emergence of evidence-based medicine; and the fragmented development of public health as a discipline. It is argued that modern air pollution epidemiology reflects a prevalent shallow environmentalism.

6

Measuring the health effects of polluted air: quantitative risk assessment case study

In this chapter a case study is introduced, which then forms the basis of exploration of the philosophical aspects of developments in modern epidemiology in the following two chapters. The case study involves a relatively modern public health technique called quantitative risk assessment (QRA). In a two-year piece of work between 1997 and 1999 a QRA was used to assess the impact of air pollution on the health of a population in a region of north-east Thames. At that time, the technique had not, to my knowledge, been used previously at the local level, although it had been used by the UK Department of Health in 1998 to assess the impact of air pollution nationally, as well as in a few other cities globally in the late 1990s.

The findings of this QRA have already been published [1] and, although it will be necessary to summarise these, the aim of this chapter is instead to introduce the case study as a vehicle through which to later investigate some of the problems of public health theory and practice today. After a brief description of background developments, this chapter outlines the QRA undertaken, and then looks at the scientific limitations of the technique. Although this book focuses on outdoor air pollution, a short section on indoor air emphasises the significance of this kind of pollution, especially for low-income countries. Philosophical and ethical considerations are then examined in Chapters 7 and 8.

The environment, air pollution and public health: international and national contexts 1970–98

As the epidemic shift from acute infectious diseases to non-communicable diseases in high-income populations became established by the second half of the twentieth century [2, 3], awareness of the damaging aspects of Western lifestyles mounted. As covered earlier, concerns about the impact of industrialisation and urbanisation on the environment germinated in the nineteenth century but they resurfaced around the 1960s accompanied by a rethinking of humankind's relationship with the natural world [4, 5], and alongside sometimes radical reappraisals of the role and limitations of scientific medicine [6–8].

In the international policy arena the changing social climate was reflected at the 1972 United Nations (UN) Environment Conference. World Health Organization (WHO) interest in air pollution impacts and monitoring continued [9, 10], then in 1987 the UN published *Our Common Future*, prepared by the World Commission on Environment and Development, a document that identified several trends in environmental deterioration and also discussed in detail for the first time the now widely used term 'sustainable development' [11].

In 1992 the WHO held a Commission on Health and the Environment, the same year as the landmark Earth Summit in Rio de Janeiro, Brazil. At Rio the signing of 'Agenda 21' was significant in stating clearly the need for integration of environmental, economic and social planning at national and international levels. To follow up on the implementation of Agenda 21, the UN Commission on Sustainable Development was created in 1996.

Within this changing atmosphere it was not surprising to see a reorientation of ideas towards how the physical environment, including air, might affect human well-being. Increase in attention given to environmental health in the UK was evident in the 1992 White Paper *Health of the Nation*, which stressed a 'growing acceptance of responsibility for the quality of the environment' as well as 'an understanding that the efforts of individuals are as important to the creation of a healthy environment as the actions of the Government' [12]. In 1994 the Government's *Sustainable Development* strategy made clear its commitment to environmental protection [13].

In 1996 the Government (Department of the Environment) published the significant *United Kingdom National Environmental Health Action Plan* (NEHAP). This document defined environmental health as 'those aspects of human health, including quality of life, that are determined by physical, biological, social and psychosocial factors in the environment' and refers to the theory and practice of assessing, correcting, controlling and preventing those factors in the environment that can potentially affect adversely the health of present and future generations' [14].

One section of the NEHAP deals specifically with air quality, and informs the reader of the imminent *UK National Air Quality Strategy (NAQS)*, which duly arrived in 1997 in part fulfilment of Part IV of the 1995 Environment Act: Local Air Quality Management (LAQM) [15]. The NAQS described a comprehensive approach to maintaining and improving the quality of ambient outdoor air in the UK. It set 'health-based air quality standards, air quality objectives which it is intended should be achieved by the end of 2005, and the process by which those objectives will be achieved' [16]. Although the NAQS acknowledges the importance of national policies in improving air quality, it places significant responsibilities with local authorities and promises resources to assist.

So, within these national and international contexts, an opportunity existed to undertake a unique piece of practical, service-based public health work, involving quantitative application of epidemiological data. It was deemed highly relevant to do so for two further reasons.

First, if public health is considered 'the science and art of preventing disease, prolonging life and promoting health through organised efforts of society' [17], then alongside the growing awareness of the health effects of air pollution comes a responsibility for public health, with its particular scientific and allied skills, to attempt to tackle a multifaceted health issue.

Second, the causes and effects of air pollution mean that strategies needed to tackle air pollution are cross-disciplinary, requiring the kind of multisectoral approach encouraged in the 1997 White Paper *The New NHS* [18], the 1998 Green Paper *Our Healthier Nation* [19] and strongly advocated by Sir Kenneth Calman in the 1998 *Chief Medical Officer's Project to Strengthen the Public Health Function in England* [20]. Public health would appear to be ideally placed to take a central role in the organised efforts needed to address a complex health problem such as this.

Quantitative risk assessment

The QRA (background methodology, data gathering, process and calculations) will be presented here in outline only, focusing on what is salient to this book.

Background methodology: time-series studies in epidemiology

The QRA method was developed initially in relation to environmental cancer risks. It is an applied procedure that uses literature-derived dose-specific risk estimates to predict the health impact of some specified (usually population-based) distribution of exposure to an identifiable factor [21]. In air pollution QRA the risk estimates are predominantly derived from time-series studies so, before looking at the QRA itself, it is necessary to look at the preceding step.

The time-series analysis (or study) is an epidemiological method designed to assess, for example, associations between health indicators such as mortality and levels of air pollution that are not perceived as severe, but fall within existing air quality standards or guidelines [22–24]. In other words, time-series studies can look at the health effects of mild diurnal fluctuations in air pollution, rather than focusing on the effects of acute severe episodes such as smogs.

More specifically, time-series studies examine the relationship between an exposure variable, such as air pollutant(s), and an outcome variable, such as deaths over the same time units (usually days), for a period of usually one or more years [25, 26]. The end-product of a time-series analysis is a regression coefficient of the association of one variable with another.[1] For example, Schwartz found – over an

11-year period in Steubenville, Ohio – that an increase in total daily particulates of 100 $\mu g/m^3$ was significantly associated with a 4% increase in mortality on the succeeding day [27].

There have now been many such studies performed in various geographical locations, looking at different air pollutants and different health outcomes. Importantly, it is the pool of these study results that informs QRA [28] and a review of the literature suggested that the four major outdoor air pollutants – particulate matter (PM_{10}), sulphur dioxide (SO_2), nitrogen oxides (NO_x) and ozone (O_3) – are associated with adverse health outcomes.[2] Increasing levels of the pollutants are associated with an increase in deaths (from all causes) and an increase in respiratory hospital admissions. The nature and source of these pollutants in London and the UK are shown in Table 6.1. The review of the evidence of the health effects can be found in the original publication [1].

The method

There have been several descriptions of methods for air pollution QRA [29]. This QRA followed, as closely as possible, those described in the 1998 publication by the Department of Health Committee on the Medical Effects of Air Pollution (COMEAP) entitled *Quantification of the Effects of Air Pollution on Health in the United Kingdom* [30]. There are four essential components [31]:

1. *Identification of a health hazard.* The QRA focused on four outdoor air pollutants (PM_{10}, SO_2, NO_x and O_3) and two health outcomes – deaths (all-cause) and respiratory hospital admissions.
2. *Definition of dose–response relationship.* The same exposure–response (risk) coefficients were used as those worked out by COMEAP.
3. *Estimation of the population's profile of exposure to the health hazard.* Data from local air pollution monitors were used when possible, and data from the nearest available monitor otherwise. Unlike COMEAP, Emission Inventories or a Geographical Information System were not employed, as the aim was to demonstrate what is available and possible locally. The approach was considered scientifically reasonable given the size and geography of the region.
4. *Estimation of aggregate additional health risk attributable to that exposure profile.* The number of theoretically preventable deaths and respiratory hospital admissions caused by air pollution were calculated, as well as the total of those attributable to air pollution.

The location and data collected

Barking and Havering Health Authority (BHHA)[3] is situated at the north-east corner of the London region. BHHA comprised, at the time, two local authorities: London Borough of Barking and Dagenham (LBBD) and London Borough of

Table 6.1 Sources of atmospheric pollutants in London [54] and the UK [55]

Pollutant	Source (London)	Source (UK)
Particulate matter (PM$_{10}$) PM$_{10}$ has a primary component emitted directly from sources such as road traffic, a secondary component formed by reactions of atmospheric gases, and a coarse component arising from surface soils, dusts and sea spray [56].	Road traffic (77%) Industry (11%) Waste treatment and disposal (5%) Air traffic and trains (5%) Domestic/commercial (2%)	Industry (58%) Road transport (26%) Domestic/commercial (13%) Other (3%)
Nitrogen oxides (NO$_x$) NO$_x$ is a collective term describing nitric oxide (NO) and nitrogen dioxide (NO$_2$). Most urban NO$_2$ is derived from the interaction of NO with atmospheric oxygen or ozone [57]. NO is formed mainly when fuels are burnt [58].	Road traffic (75%) Air traffic and trains (7%) Industry (6%) Waste treatment and disposal (1%)	Transport (56%) Industry (39%) Domestic (3%) Other (2%)
Sulphur dioxide (SO$_2$) SO$_2$ is produced from the burning of fossil fuels and was largely responsible for the 'smog' episodes in the first half of this century. Concentrations have steadily fallen with the decline in use of domestic coal [59].	Industry (59%) Road traffic (23%) Air traffic and trains (6%) Waste treatment and disposal (2%)	Power stations (66%) Other industry (24%) Transport (4%) Domestic (3%) Other (3%)
Ozone (O$_3$) In the lower atmosphere ozone is a secondary pollutant formed mainly from NO$_x$ and hydrocarbons in the presence of sunlight [60]. Rural levels tend to be higher because of scavenging of ozone by nitric oxide in urban areas [61].	*Secondary pollutant* Precursors arise mainly from motor vehicles and industry.	*Secondary pollutant* Precursors arise mainly from motor vehicles and industry.

Havering (LBH). Located in the south-west part of the health authority the urban and more deprived LBBD is geographically smaller than LBH, has less people, more industry and lower car ownership (Table 6.2).

Through liaison with Environmental Health Officers data on air pollutant levels were obtained from records of monitoring equipment owned by the two boroughs. When this was unavailable data from the nearest London monitoring station were used instead. Averaging of pollutant data for the presence of more than one local monitor was performed according to the protocol of a large European study [32].

Table 6.2 Comparison of boroughs in Barking and Havering Health Authority by factors relevant to air pollution 1997

	London Borough of Barking and Dagenham (LBBD)	London Borough of Havering (LBH)
Area	34 km² (23% of health authority)	112 km² (77% of health authority)
Population	153 715 (40% of health authority)	230 909 (60% of health authority)
Ordinance Survey	Urban (>20% of land covered)	Rural (<20% of land covered)
Deprivation[a]	Townsend score: +3.5	Townsend score: −2.4
Smoking prevalence	Male 26%; female 34% (1994)	Male 24%; Female 27% (1994)
Industrial processes[b]	10 Part A; 26 Part B	1 Part A; 15 Part B
Car ownership	56.2% households with car (1991)	74.2% households with car (1991)

[a] A positive Townsend score indicates above-average deprivation, a negative Townsend score indicates below average. Deprivation increases with higher positive values and decreases with lower negative values.
[b] Part A industrial processes are generally more complex ones, monitored by the Environment Agency. Part B industrial processes are monitored by local authorities.

The two health outcomes used in the air pollution QRA were all-cause mortality and respiratory hospital admissions, and data for these were obtained from the Public Health Common Data Set and the Hospital Episode Statistics (HES) respectively [1].

The process and final results

The stages involved in the QRA process were as follows:

1. *Annualisation values.* Annualisation values (annual mean of 8-hour or 24-hour means) were calculated for each pollutant for each year with available data.
2. *Calculation.* The annualisation value for a given year was applied to the baseline health outcome data for that year using the appropriate risk coefficient.
3. *Attributable vs preventable health outcomes.* The initial calculation above provided the number of hospital admissions or deaths *attributable* to an air pollutant. However, by not using a 'threshold' of effect, this method of analysis attributes health effects from a baseline air pollutant level of zero. Reduction of air pollution levels to zero is not realistically achievable so, in addition, *preventable* deaths and hospital admissions were calculated by only accounting for those that occurred above a local threshold (defined as the lowest value for the pollutant in that year). In terms of current public health practice, preventable outcomes are more relevant and so have been presented as the final estimates.
4. *Sensitivity analyses.* Sensitivity analyses were performed on all calculations (to account for capture rate of the pollutant monitors of between 80% and 90%) and, in general, the ranges were reassuringly narrow.

Table 6.3 Total preventable deaths (all-cause) and respiratory hospital admissions in Barking and Havering Health Authority caused by the major air pollutants 1997

Pollutant	Deaths (1997)	Respiratory hospital admissions (1997)
Ozone (O_3)	76.5	71.9
Particulates (PM_{10})	49.6	42.6
Sulphur dioxide (SO_2)	(54.3)[a]	(38.8)[a]
Nitrogen dioxide (NO_2)	–	72.5
Total	180.4	225.8

[a] 1994 value as proxy.

5. *Final results.* Piecing together the calculations performed provided one final table for the whole health authority (Table 6.3), showing deaths and respiratory hospital admissions attributable to air pollution and preventable. Although there were difficulties with data collection in most years, *the final estimates produced by the QRA showed that in BHHA in 1997 there were approximately 180 deaths and 226 respiratory hospital admissions attributable to air pollution, and theoretically preventable.*

Overall, the results of the QRA suggest that existing levels of air pollution have significant effects on mortality and morbidity at the local level [33]. In BHHA – a district with about 385 000 residents – the 180 preventable deaths in 1997 attributable to air pollution made up just over 4% of the total 4404 deaths for that year. This compares with 44 deaths from breast cancer and 30 deaths from diabetes mellitus in BHHA in 1996 [34]. The impact of air pollution will be mostly on individuals with existing chronic respiratory disease, and is likely to be greater in LBBD than in LBH (inner city, higher population density, higher deprivation, higher concentration of industry, higher smoking rates, higher death rates and higher respiratory hospital admission rates) although this disparity was not strikingly evident due to the limitations of available local pollutant data.

Scientific considerations: technical and methodological limitations

Quantitative risk assessment is an approach of particular interest to health policy-makers because it produces numerical estimates of health impact which can be used when weighing up competing health concerns. There are, however, limitations to the QRA approach which mean the results should be treated carefully.

The earliest methodological problems along the QRA pathway relate to exposure measurement and are technical. The measurement of air pollution is a developing science and the last few years have witnessed changes in technology and technique.

Monitors may be part of a national network or locally operated, automatic (regular measurements fed back via telephone lines) or non-automatic (visits required to collect data) and able to measure one pollutant or several. Inconsistency in equipment used, accuracy of measurement, and monitoring sites chosen all place limits on the validity of data obtained [35–37].

Next are methodological concerns. Ambient levels of pollutants are actually only proxy measures of true population exposure; the annualization procedure is really a default option in the absence of more accurate exposure data. Also, the QRA treated the health authority population as if evenly exposed to the levels of pollutants described by nearby fixed-site monitors. In reality this is unlikely to be the case [38].

Problems also exist with regard to the risk coefficients. First is the issue of appropriateness of application of coefficients calculated in a study on one population to another population. Second, the presumed relationship between pollutant concentration and health outcome may not be linear, although this study took it to be [39].

Other methodological limitations relate more directly to interpretation of the data. Many of the deaths attributed to air pollution under the algorithm used in this analysis are likely to entail only minor bringing forward in time rather than being new deaths (i.e. otherwise unexpected). This effect has been termed 'harvesting'. But even if the excess deaths represent the deaths of already ill people being brought forward, such a period of months or years is still of public health importance [40].

It is also difficult to separate out the effects of air pollutants. In using single-pollutant models there may be double-counting of deaths causing over-estimation of effect. Along similar lines, it is hard to discern whether hospital admissions are repeat events on the same individual or on different people, although this does not necessarily diminish the significance of the event [41].

In contrast to concerns about over-estimation, it could be argued that the impact of air pollution on health is likely to be under-estimated by the QRA because time-series analyses tell us only about the acute effects of increased pollutant levels and tell us nothing about the long-term effects of air pollution on the incidence of chronic diseases [42, 43].

Finally, the QRA undertaken in BHHA looked only at two health outcomes, whereas the true impact of air pollution is likely to be substantially higher including additional hospital admissions for cardiovascular disease [44], additional attendances at accident and emergency departments and general practices, costs of these additional attendances and hospital admissions, and the costs of additional drug prescribing.

With an outline of the QRA process described, along with its scientific strengths and weaknesses, the following two chapters will look at the broader framework within which the QRA is shaped, and some of the philosophical, ethical and practical issues. First, however, it is necessary to look at indoor air pollution, and how this differs from air pollution outdoors, in terms of causation and impact.

Indoor air

The theme of this book is atmospheric air, and the changing conceptions of (atmospheric) air and health. In contemporary epidemiological and environmental health literature, atmospheric air may also be referred to as ambient or outdoor air, in other words air outside of buildings. This differentiates from indoor air, and the study of indoor air pollution.

Over the past few decades epidemiological research has tended to focus on outdoor air pollution [45]. As is explored in Part III, outdoor air pollutant concentrations have generally fallen in Western countries through the second half of the twentieth century, although health effects have been more recently demonstrated for smaller size particles and at concentrations lower than previously anticipated.

Outdoors and indoors are in one sense connected, in that what is present in the air outside buildings may permeate inside. Outdoor air pollution is indeed a significant source of pollution indoors, and this underpins the advice during high air pollution episodes for those with chronic respiratory diseases to close their windows and doors. Indoor and outdoor air pollution are also linked in that the epidemiological methods to investigate their health effects overlap [46]. Significantly, for both types of pollution epidemiological research has focused on the concerns of Western, or high-income, countries. This is reflected in how indoor air pollutants have been traditionally grouped [47]:

- combustion products, e.g. particulates from cooking, and environmental tobacco smoke;
- building materials, furnishings, and chemical products, e.g. semi-volatile and volatile organic compounds, or asbestos;
- pollutants in the ground underneath buildings, e.g. radon in soil;
- pollutants from biological processes, e.g. microbes, mites and other allergens.

As can be seen from the above list, indoor air pollutants differ from outdoor pollutants in their origins and health implications, as well as with respect to issues of poverty and public health. For Western societies, the health effects relating to the groupings above are predominantly respiratory, and are relatively moderate (aside from the cancers associated with asbestos, tobacco smoke, radon or combinations of these) [48]. On a global scale, however, the health issues for Western countries are overshadowed by the health concerns arising from indoor air pollutants in low-income countries [49].

In low-income countries the burning of solid fuels for heating and cooking is by far the greatest source of air pollution indoors, and causes levels of air pollutants, such as particulates, that are rarely experienced in Western countries. The use of biomass (any material derived from plants or animals and deliberately burnt by humans) for cooking and heating is directly related to poverty, with approximately half the world's population, and up to 90% of rural households in developing countries, relying on unprocessed biomass fuels. The relationship of fuel use to poverty is captured by the image of the 'energy ladder'. As one progresses up the ladder the energy-generating resources that are used tend to be cleaner and have increased efficiency in terms of combustion: animal dung is on the lowest rung of the ladder, succeeded by crop residues, wood (the most commonly used form of biomass), charcoal, kerosene, gas and electricity [50].

Levels of indoor air pollutants are affected by factors such as source of pollutant, combustion equipment and process, and ventilation of the indoor space. In low-income countries the materials used for combustion are in raw form, the apparatus employed may be little more than piles of rocks as an open fireplace, and ventilation is often unpredictable. As a result, levels of indoor pollutants are often many times higher than recommended in international air quality guidelines. Duration of exposure is also important, and the concept of 'micro-environments' has been created to address the changing nature of individuals' exposure over time (for instance during the course of a day) [51]. Globally, people spend a substantial proportion of their time indoors, but in low-income countries women and young children are particularly vulnerable to indoor air pollution as their levels of exposure are often higher due to domestic cooking responsibilities and child-rearing. This is especially marked in rural areas.

Exposure to indoor air pollution in low-income countries has been associated with health effects that include respiratory disease (childhood acute respiratory infections, chronic obstructive pulmonary disease, asthma, tuberculosis), cancers (lung, nasopharyngeal, laryngeal), eye problems and blindness (cataract), and low birthweight and infant mortality [50]. In terms of global burden of disease, it has been estimated that indoor air pollution may be responsible for 4–5% of the global low-income countries' totals for deaths and disability-adjusted life-years lost (DALYs), comparable with tobacco use and unsafe sex. The largest single category is acute respiratory infections in children under five, representing 64% of deaths and 81% of DALYs, and estimated to be responsible for 1.2 million premature deaths annually in the early 1990s [45].

The policy response to the problem of indoor air pollution has been inadequate. To date, 'the total money spent globally on all national and international efforts to find and implement solutions is less than the cost of pollution controls on one typical new power plant' [48]. There has been a call for much stronger investment in epidemiological research around the causes and health effects of indoor air pollution in low-income countries, as well as development and evaluation of interventions that reduce exposure, such as improved cooking stoves [52].

Unfortunately, however, there may be an underlying assumption that the solution simply lies in industrial development, and progression to energy-generating resources higher up the ladder. While this position may have some historical resonance, it is lamentable in terms of the damage currently experienced by those worst off in the world. The neglect of indoor air pollution as an issue in global public health is considered by some as a gross injustice, especially in relation to particular groups [53]:

. . . few if any large groups are more disenfranchised and disadvantaged than poor rural women in developing countries and their young children, who experience the bulk of global airborne exposures to many pollutants.

The ideas of global and environmental justice are returned to in Part IV of this book, and a brief look at indoor air pollution has laid out some important issues. It is necessary now, however, to return to the theme of Part III, and the following chapter continues to look philosophically at the QRA case study.

NOTES

1. Studies on links between air pollutants and morbidity and mortality, such as analyses of the effects of the London smogs, precede time-series analyses and are covered in Part II.
2. The literature review involved a search strategy including on-line databases, publications from specialist centres and Government departments, hand searching of the *Index Medicus* and other key journals, and guidance towards articles from peers.
3. Following a restructuring of the NHS in 2002 BHHA ceased to exist. In its place Barking and Dagenham Primary Care Trust (PCT) and Havering PCT were created.

REFERENCES

1. A. S. Kessel, A. J. McMichael and C. J. Watts, Quantitative risk assessment of the impact of air pollution in Barking and Havering Health Authority. *Pub. Health Med.*, **2**:1 (2000), 13–19.
2. A. R. Omran, The epidemiologic transition. A theory of the epidemiology of population change. *Milbank Mem. Fund. Q.* **49** (1971), 509–38.
3. D. Fox, *Power and Illness* (Berkeley: University of California Press, 1993).
4. R. Carson, *Silent Spring* (New York: Houghton Mifflin, 1962).
5. J. R. Des Jardins, *Environmental Ethics* (New York: Wadsworth, 1997).
6. R. Dubos, *Mirage of Health* (London: George Allen & Unwin Ltd, 1960).

7. I. Illich, *Limits to Medicine* (London: Marion Boyars, 1976).

8. T. McKeown, *The Role of Medicine* (Oxford: Blackwell, 1979).

9. Anonymous, The long-term effects of air pollution on health. *WHO Chronicle*, **28** (1974), 12–15.

10. G. Akland, J. Kretzschamer, D. Fair and H. de Konig, Air quality surveillance: trends in selected urban areas. *WHO Chronicle*, **34** (1980), 147–52.

11. D. Brown, The need to face conflicts between rich and poor nations. In L. Westra and H. Werhane, eds., *The Business of Consumption* (Lanham: Rowman & Littlefield, 1998).

12. Department of Health, *The Health of the Nation* (London: HMSO, 1992).

13. —, *Sustainable Development: The UK strategy.* (London: HMSO, Cm2426, 1994).

14. Department of the Environment, *The United Kingdom National Environmental Health Action Plan* (London: HMSO, 1996), p. 2.

15. Department of the Environment, Transport and the Regions, *The United Kingdom National Air Quality Strategy and Local Air Quality Management: Guidance for local authorities*, Environmental Circular 15/97 (London: HMSO, 1997).

16. *Ibid.*, p. 2.

17. Faculty of Public Health Medicine (UK), *Handbook on Education and Training* (London: FPHM, 1998).

18. Department of Health, *The New NHS: Modern, dependable* (London: HMSO, 1997).

19. —, *Our Healthier Nation* (London: HMSO, 1998).

20. —, *Chief Medical Officer's Project to Strengthen the Public Health Function in England* (London: HMSO, 1998).

21. B. Ostro, *Estimating the Health Effects of Air Pollutants: A method with an application to Jakarta* (Washington: The World Bank, Policy Research Working Paper 1301, 1994).

22. B. Brunekreef, D. Dockery and M. Krzyanowski, Epidemiologic studies on short-term effects of low levels of major ambient air pollution components. *Env. Health Persp.* **103**:S2 (1997), 3–13.

23. K. Katsouyanni, G. Touloumi, C. Spix *et al.* Short-term effects of ambient sulphur dioxide and particulate matter on mortality in 12 European cities: results from time series data from the APHEA project. *BMJ*, **314** (1997), 1658–63.

24. Department of Health (Committee on the Medical Effects of Air Pollution), *Handbook on Air Pollution and Health* (London: HMSO, 1997).

25. M. Lebowitz, Epidemiological studies of the respiratory effects of air pollution. *Eur. Resp. J.* **9** (1996), 1029–54.

26. K. Katsouyanni, Research methods in air pollution epidemiology. In T. Fletcher and A. J. McMichael, eds., *Health at the Cross-roads: Transport policy and urban health* (Chichester: John Wiley & Sons, 1997), pp. 51–60.

27. J. Schwartz and D. Dockery, Particulate air pollution and daily mortality in Steubenville, Ohio. *Am. J. Epidemiol.*, **135** (1992), 12–20.

28. J. Schwartz, C. Spix, G. Touloumi *et al.*, Methodological issues in studies of air pollution and daily counts of deaths or hospital admissions. *J. Epidemiol. Comm. Health*, **50**:Suppl. 1 (1996), S3–S11.

29. B. Ostro, *A Methodology for Estimating Air Pollution Health Effects* (Geneva: World Health Organization, WHO/EHG/96.5, 1996).

30. Department of Health (Committee on the Medical Effects of Air Pollution), *Quantification of the Effects of Air Pollution on Health in the United Kingdom* (London: HMSO, 1998).

31. London Review Group, *Review of Methods Proposed, and Used, for Estimating the Population Exposure to Urban Air Pollution* (London: Report for the World Bank, 1995).

32. K. Katsouyanni, G. Touloumi, C. Spix *et al.*, Short-term effects of ambient sulphur dioxide and particulate matter on mortality in 12 European cities: results from time series data from the APHEA project. *BMJ*, **314** (1997), 1658–63.

33. B. Brunekreef, D. Dockery and M. Krzyanowski, Epidemiologic studies on short-term effects of low levels of major ambient air pollution components. *Env. Health Persp.* **103**:S2 (1997), 3–13.

34. Barking and Havering Health Authority, *Statistical Data Accompanying the Annual Report of the Director of Public Health 1997* (London: Barking and Havering Health Authority, 1997).

35. Department of the Environment, Transport and the Regions, *Monitoring for Air Quality Reviews and Assessments: Appendix 3 – definition of site classes* (London; The Stationery Office, 1998), pp. 70–2.

36. —, *Monitoring for Air Quality Reviews and Assessments* (London: The Stationery Office, 1998).

37. —, (Report of the Airborne Particles Expert Committee), *Source Apportionment of Airborne Particulate Matter in the United Kingdom* (London: HMSO, 1999), p. ii.

38. A. D. McDonald, P. F. Ricci, T. Beer and A. A. Moghissi, Application of modelling and simulation in risk analysis. *Env. Intl*, **25**:6/7 (1999), 691–2.

39. T. Beer and P. F. Ricci, A quantitative risk assessment based on population and exposure distributions using Australian air quality data. *Env. Intl*, **25**:6/7 (1999), 887–98.

40. S. Zeger, F. Dominici and J. Samet, Harvesting-resistant estimates of air pollution effects on mortality. *Epidemiology*, **10** (1999), 171–5.

41. K. Rothman and S. Greenland, *Modern Epidemiology* (Philadelphia: Lippincott-Raven, 1998).

42. A. J. McMichael, H. R. Anderson, B. Brunekreef and A. J. Cohen, Inappropriate use of daily mortality analyses to estimate longer-term mortality effects of air pollution. *Int. J. Epidemiol.*, **27** (1998), 450–3.

43. B. Brunekreef, Air pollution and life expectancy: is there an association? *Occ. Env. Med.*, **54** (1997), 781–4.

44. T. P. Brown, L. Rushton, M. A. Mugglestone and D. F. Meechan, Health effects of a sulphur dioxide air pollution episode. *J. Pub. Health Med.*, **25**:4 (2003), 369–71.

45. K. R. Smith and S. Mehta, The burden from indoor air pollution in developing countries: comparison of estimates. *Int. J. Hyg. Env. Health*, **206** (2003), 279–89.

46. J. Samet, Environmental and occupational health sciences in public health. In R. Detels, J. McEwen, R. Beaglehole and H. Tanaka, eds., *Oxford Textbook of Public Health* (Oxford: Oxford University Press, 2004), pp. 959–78.

47. B. H. Chen, C. J. Hong, M. R. Pandey and K. R. Smith, Indoor air pollution in developing countries. *World Health Stat. Quart.*, **43** (1990), 127–38.

48. K. Smith and J. Sundell, Editorial. *Indoor Air*, **12** (2002), 145–6.

49. A. J. McMichael, *Human Frontiers, Environments and Disease: Past patterns, uncertain futures* (Cambridge: Cambridge University Press, 2001), pp. 172–4.

50. N. Bruce, R. Perez-Padilla and R. Albalak, Indoor air pollution in developing countries: a major environmental and public health challenge. *Bull. World Health Org.*, **78**:9 (2000), 1078–92.

51. K. R. Smith, J. M. Samet, I. Romieu and N. Bruce, Indoor air pollution in developing countries and acute lower respiratory infections in children. *Thorax*, **55** (2000), 518–32.

52. R. Smith, Indoor air pollution in developing countries: recommendations for research. *Indoor Air*, **12** (2002), 198–207.

53. K. R. Smith, J. M. Samet, I. Romieu and N. Bruce, Indoor air pollution in developing countries and acute lower respiratory infections in children. *Thorax*, **55** (2000), 530.

54. London Research Centre, *London Atmospheric Emissions Inventory* (London: London Research Centre, 1997).

55. Department of the Environment, Transport and the Regions (National Environment Technology Centre), *National Atmospheric Emissions Inventory*. www.naei.org.uk (accessed May 2005).

56. —, (Report of the Airborne Particles Expert Group), *Source Apportionment of Airborne Particulate Matter in the United Kingdom* (London: DETR, 1999).

57. Department of the Environment (Expert Panel on Air Quality Standards), *Nitrogen Dioxide* (London: HMSO, 1996).

58. London Research Centre, *London Atmospheric Emissions Inventory* (London: London Research Centre, 1997), pp. 51–2.

59. Department of the Environment (Expert Panel on Air Quality Standards), *Sulphur Dioxide* (London: HMSO, 1995).

60. B. G. Higgins, H. C. Francis, C. J. Yates *et al.*, Effects of air pollution on symptoms and peak expiratory flow measurements in subjects with obstructive airways disease. *Thorax*, **50** (1995), 149–55.

61. Department of the Environment (Expert Panel on Air Quality Standards), *Ozone* (London: HMSO, 1994).

7

Epidemiological theory and philosophical considerations

The quantitative risk assessment (QRA) presented in the previous chapter has two fundamental components: the time-series epidemiological studies that determine the risk coefficients; and application of these coefficients to enumerate in a given population the impact of air pollutants. While technological and methodological problems have been looked at in the previous section, to better understand QRA it is important to explore how the constituent components fit into the broader picture of developments in epidemiological theory and public health philosophy and practice.

This process is two-way since an appreciation of how QRA 'fits in' is correspondingly informative of difficulties in contemporary epidemiological and public health thinking. The chapter is subdivided for convenience and obvious links exist between the sections.

Historical developments in epidemiology

Epidemiology is difficult to define, in part because what is considered to be modern epidemiology differs markedly from earlier conceptions. The *Shorter Oxford English Dictionary on Historical Principles* places the first appearance of the word in 1873 and describes epidemiology succinctly as 'that branch of medical science which treats of epidemics' [1]. However, a well-known classroom epidemiology textbook of today chooses to draw on a 'useful and comprehensive' 1970 definition of epidemiology as 'the study of the distribution and determinants of disease frequency', to which the phrase 'in human populations' is added [2]. Tellingly, a 1994 edition of the highly regarded *Essential Public Health Medicine*[1] limits its definition of epidemiology to 'one of the population sciences basic to public health' [3].

What these three varying definitions reveal is both the developmental transitions through which epidemiology has passed, and some of the confusion over what constitutes epidemiological theory today. As seen in Part I, when epidemiology emerged in the late nineteenth century, it was largely a descriptive enterprise.

Drawing on demography, and using basic statistics, medical scientists described differences in disease patterns between population groups. The incidence of diseases in communities could be compared classically by area but also by parameters such as age, sex and occupation. Although the term 'epidemic' conjures up an image of infectious diseases, in fact it can relate – as the *Shorter Oxford English Dictionary on Historical Principles* again informs us – to any disease 'prevalent among a people or community at a special time, and produced by some special causes not generally present in the affected locality' [1].

Based on observed differences in diseases between communities, medical scientists speculated on disease aetiologies. So, for instance, noticing differences between urban and rural communities in rates of respiratory diseases, such as pneumonia, led to the belief that smoke pollution might be responsible. This was reinforced by statistics suggesting rates increased in cities after episodes of smog. In charting broad historical developments in epidemiology Susser and Susser have labelled this period 'the era of sanitary statistics', with its paradigm of miasma and an analytical approach of demonstrating clustering of morbidity and mortality [4]. Like others, as discussed in Parts I and II, they link this epidemiological era to the preventative approach (at a population level) of public health of that time, characterised by efforts to improve health through improving drainage, sanitation and sewage disposal.

Towards the end of the nineteenth century, germ theory became the dominant paradigm. As the Henle–Koch postulates[2] were accepted, the scientific community began to search for single agents responsible for specific diseases [5]. This era of infectious-disease epidemiology, which lasted well into the twentieth century, focused on laboratory culture from disease sites, experimental transmission of diseases and reproduction of lesions. The prevention of illness centred on the interruption of disease transmission through vaccines, isolation of cases and, later, antibiotics [4].

The transition between these two eras – from sanitary statistics to infectious diseases – marked an important turning-point in epidemiology, one that forms the roots of some of the problems in current epidemiological thought. Whereas very early epidemiology was interested in examining the basic social and biological factors that could explain community health and ill-health, a gradual shift occurred towards investigating specific factors associated with particular diseases. In tandem with this there was a move from thinking about what makes populations ill to what makes individuals ill [6]. Within public health a biomedical model of disease was eroding a social model of population health.

For Western medicine, models of human disease aetiology had a certain linear progression in the second half of the twentieth century. As the impact of chronic diseases began to overshadow that of acute infectious disease in Western countries, models incorporated the relevance of the host and the environment – in addition to the agent – in disease causation, and later introduced the importance of bio-psychosocial factors. But this backdrop created a state of uncertainty for a population science, and in the 1950s and 1960s epidemiology found itself at something of a crossroads.

One reaction to these developments was 'social medicine', defined by its first professor, John Ryle, in the 1940s as a sociobiology of health and disease, grounded in holistic epistemology and a deep rejection of mechanistic positivism. The historian Dorothy Porter argues that, in actuality, social medicine referred to a disputed range of linked concepts: in one sphere it was a scientific humanism directed towards deconstructing bedside medicine and replacing specialists with practitioners of whole-person medicine; and, in another, it represented an academic endeavour to explain, through quantitative science, the social basis of illness in populations [7]. Social medicine will be discussed in Chapter 8 but, importantly, it illustrates the path epidemiology and public health could have taken, back toward their origins.

Epidemiology since the 1960s

Instead epidemiology went in a different direction, one that paralleled the reductionistic leanings of Western medicine. From the 1960s onwards epidemiology increasingly focused on examining the relationship, or association, between an exposure and an outcome. Variously termed 'risk-factor epidemiology', 'analytical epidemiology' or 'modern epidemiology'[3] all refer to the academic practice of using epidemiological methods to measure putative associations between ever more specific risk factors to which groups may be exposed, and health outcomes in those groups [8].

Here it is possible to see how the time-series studies that inform the QRA fit in. Each study is basically an investigation of the relationship between the local atmospheric concentration of specific air pollutants (the exposure) and the amount of death or hospital admission for respiratory disease in that locality (the outcome). For the exposure, the pollutant researched has historically been a progressively more reduced version of that previously examined. For instance, originally smoke was investigated, then this became black smoke, then total suspended particulates, then particulates less than 10 µm in diameter, and the latest research indicates that particulates less than 2.5 µm in diameter are those most relevant. To these, other specific pollutants – nitrogen dioxide, sulphur dioxide and ozone – have gradually been added. As was described in Chapter 6, one of the public health responses to these studies, such as the National Air Quality Strategy Standards and Objectives (UK), specifies desired levels of these pollutants in the atmosphere.

This contemporary period has been coined 'the era of chronic disease epidemiology', with its paradigm 'the black box'. Sometimes simply referred to as 'black-box epidemiology', the phrase represents a belief in the discipline that what goes on inside the box is of secondary importance to associations found between the factors either side of the box; in other words, it is a method that can suggest adverse effects of exposures without a full understanding of the causal pathway [9].

Black box epidemiological research is risk-factor epidemiology with an emphasis on searching for environmental causes of disease, born out of a belief in the 1980s that perhaps 80% of cancers might be caused by environmental factors [10, 11]. An investigation in black-box epidemiology such as a cohort study usually provides a summary statistic that compares the likelihood of an outcome between those exposed (to a putative risk factor) and those unexposed, called the relative risk (RR). Despite its name the RR can reveal the exposure as a protective factor – such as exercise for coronary heart disease – or a risk factor, such as smoking or an atmospheric pollutant for lung disease. With epidemiological studies that measure exposure and outcome on individuals (e.g. does an individual smoke, and has that individual developed coronary heart disease?) the RR informs of risk to the individual, and tallies with the individual preventive approach of much modern public health work, for instance through advocating lifestyle change.

The time-series studies informing the QRA, however, measure exposure and outcome repeatedly over time at the population level and the resulting product, the risk coefficient, tells something about the risk to a population of any given level of exposure to a pollutant. The low hierarchical place of these 'correlational studies' for advocates of evidence-based medicine is looked at later in this chapter but, with the above outline in mind, it is important first to explore some of the conceptual limitations of black-box epidemiology and, by extension, of time-series studies and the QRA.

Limitations of modern epidemiology

One of the major criticisms of risk-factor or black-box epidemiology, and that from which other criticisms stem, is that there is little in the way of theory underpinning it. For at least 30 years there has been a huge effort in improving methodological technique, through more and more sophisticated ways of minimising the possibility that an association found could be explained by errors in data collection or by chance. Technological advances have enabled complex statistical modelling and multivariable analyses to facilitate this process, and risk-factor epidemiology has become a technical exercise that has concentrated on methodological refinement rather than attention to theory. One commentator reviewed 21 American textbooks and anthologies of epidemiology published between 1970 and the early

1990s, and found discussion of epidemiological theory and history to be almost non-existent [6].

A corollary of the focus on method has been the distancing of epidemiology from activities in the real world that result in improvement to public health [12]. An editor of the *American Journal of Public Health* has expressed his fears over the separation as follows [13]:

In the absence of a central concern with subject matter, the satisfactions of technical command are held within narrow bounds; in the absence of broader purpose, an arsenal of methods might not necessarily be directed to the benefit of the public health.

The times-series studies, and the QRA itself, certainly reflect this devotion to methodological precision. Most time-series studies involve large datasets, and use complex modelling and statistical analysis to find the purest possible association between an atmospheric pollutant and a health outcome. Therefore, the effects of aspects of the environment that might 'interfere' with the association are adjusted for, such as temperature, climate, seasonality, influenza outbreaks and smoking. The questionable meaning of controlling out these factors (when in the reality of population exposure they exist) will be returned to later but now serves to reflect the preoccupation with method. When QRA is undertaken it usually involves using highly sophisticated Geographical Information Systems (GIS) to plot the exposure and quantification by area, a technique inaccessible to most community-based public health workers. In an effort to reacquaint epidemiology with public health practice we opted not to use GIS in the QRA described in Chapter 6, but the process could be methodologically criticised for that reason.

Attention to method can be partly explained by the ongoing search for risk factors of ever-diminishing importance. The associations between major risk factors, such as smoking, and disease could, according to some authors, 'hardly be missed', but with most of these probably now identified researchers have been obliged over the last two or three decades to focus on much smaller associations. These, of course, require more sophisticated methods to elucidate, and often one study revealing a positive association between a risk factor and health outcome will be countered by another showing a negative association between the same factors. Hence the huge numbers of air pollution time-series studies repeatedly examining the same or very similar associations. There is a limit, some believe, to the status of black-box epidemiology posed by the number of erroneous scare stories that arise, and the public toleration of these [14].

Yet, in spite of all the methodological improvement, technological advances and the repeated studies, risk-factor epidemiology often appears not particularly helpful. This is in part, as mentioned earlier, due to its distancing from public health practice but, connected to this, risk-factor epidemiology may tell us relatively little about what makes populations ill.

Modern epidemiological research is largely underscored by a biomedical model of what makes individuals sick. Exposures have centred on factors pertaining to individuals, especially lifestyle choices (smoking, alcohol, caffeine, exercise), and outcomes have tended towards that which can be clinically detected, or has a clinical orientation (blood test results, cancer, hospital admissions). However, what makes individuals sick is not the same as what makes populations sick [15]. It is insufficient, and inappropriate, to hold that the population experience of illness simply reflects the sum of individual experiences. Although time-series studies do measure exposure and outcome at the population level, their design may strip away relevant aspects of those population experiences, thereby failing to provide integrated explanations for any association revealed. In doing their best to separate populations from their context, they tell us little about the social, political and economic factors that crucially shape population experience [16].

In addressing this, Loomis and Wing discuss whether black-box epidemiology is today's miasma theory and molecular biology is the contemporary equivalent of germ theory. While black-box epidemiology, with its search for environmental determinants, has redirected attention to an environmental theory of disease – and away from starting with clinical endpoints [3] – they lament that it has increasingly taken on the logic and research methods focused on identifying single aetiological agents and quantification of their independent effects. Both black-box epidemiology and molecular biology are misguided [17]:

The modern incarnations of these theories are centred on agents of disease that are differently conceptualized, but similarly alienated from the ecological framework from which they exist. Neither approach effectively addresses the interdependence of multiple agents or how human populations become exposed and susceptible to them. This failure derives from a view of populations as mere aggregations of individuals (i.e. vehicles for quantifying exposure–disease associations) rather than as organized groups with relational properties that cannot be deduced by measurements on individuals.

Time-series studies divorce the association between exposure and outcome from the context in which a population experiences the association in two related ways. First, the statistical controlling for confounding factors, as mentioned before, removes the effects of factors such as temperature and influenza epidemics. Yet if greenhouse warming, for example, is known to impact on temperature and infectious disease patterns, removing these confounding factors would appear to take away an important basis for understanding why and when a population may experience problems due to any given air pollutant. The danger is that 'attempting to eliminate the influence of all other causes of diseases – in an attempt to control confounding – strips away the essential historical and social context, as well as the multiple moderating influences that constitute true causation' [18].

Second, deliberately excluding social, political and economic factors from the research framework suggests that these factors are irrelevant to a population's experience of atmospheric pollution and the consequences of that exposure. For the time-series air pollution studies the following may not be considered: region/population of high deprivation with correspondingly high smoking rates; region where, historically, housing provided by the State has been built on cheap land close to industries; and low affordability of transport to get to hospital with an exacerbation of chronic respiratory disease.

It makes little sense for a science supposedly interested in what makes populations ill to remove populations in the scientific analyses from the deep-rooted factors that either make them ill or that otherwise affect their health status. But as well as making little sense, the research makes significant commitments in two related directions. First, there is a commitment to a reductionistic biomedical model of disease [19]:

... because control over measurement and extraneous factors is hindered when investigations are embedded in complex social and historical situations, this combination of influences supports the movement of the discipline away from engagement with issues of social theory, population biology and human ecology, and towards a more fundamental commitment to biomedical approaches.

Second, as Wing continues, there is an agenda that makes a political statement about the underlying reasons why populations are ill, and the actions that could be taken to improve their health [20]:

The choice is not between objective science and a science that is contaminated by social and political values. Risk factor epidemiology does not achieve objectivity by systematically examining exposure–disease associations separated from contexts of military, energy, or agriculture policy, and issues of economic inequalities and democracy. Rather, it makes a political commitment to the status quo by excluding these issues from public health consideration.

So, for QRA, what appears to be a reasonable, and scientific, way to assess the impact of atmospheric pollution on a local population may simply reflect the misguided approach of contemporary epidemiology. The time-series studies on which the QRA process is based have deep-rooted conceptual deficiencies, despite ever-improving methodological precision. Pearce [21], drawing on McKinlay, summarises the limitations:

... what is now regarded as established epidemiology is characterized by biophysiologic reductionism, absorption by biomedicine, a lack of real theory about disease causation, dichotomous thinking about disease (everyone is either healthy or sick), a maze of risk factors, confusion of observational associations with causality, dogmatism about which study designs are acceptable, and excessive repetition of studies.

A new theory? Putting the context back into epidemiology

So where does this leave the QRA? It has been argued that the time-series studies, and so by extension the QRA, are reductionistic, biomedical and contain an ingrained commitment away from the deeper social explanations and remedies for population ill health. Returning to a point made earlier, these problems reflect the distinct lack of epidemiological theory. Black-box epidemiology may be a methodological enterprise without foundations.

There have, however, been some attempts at philosophical discussion of epidemiological theory. From the late 1970s to the early 1990s there was quite extensive debate – mainly in the American academic journals – on the philosophical dimensions of scientific enquiry in epidemiology. But this was almost entirely restricted to consideration of causal inference in epidemiology, and the differences between inductive and hypothetico-deductive methods of building knowledge within the discipline [22–24]. Although some articles lamented the general lack of philosophical debate, what emerged remained strongly within the positivist tradition, for instance sometimes lengthy discussion of what Karl Popper's ideas might offer epidemiology [25–29] – with only occasional passing reference to alternative views on how science progresses and the social construction of knowledge.

What has slowly appeared in the 1990s, however, is a growing awareness that epidemiology needs to be able to understand populations, their health and disease, as entities embedded in a complex matrix of environmental, social, economic and political processes [30, 31]. The metaphor of an ecosystem has appeared. The ecosystem contains many component parts, such as population groups, which dynamically interact, and whose functions both depend on and impact on other component parts. The well-being of component parts is important but, equally, so is the health of the whole [32].

Yet despite this fresh concern about the direction of epidemiology, few serious attempts have been made to articulate a new theoretical basis. Krieger, however, has made one. She starts her cleverly titled 1994 paper 'Epidemiology and the web of causation: has anyone seen the spider?' [6] by discussing the 'web of causation', an idea first mentioned in the 1960s. Krieger complains that, although the web importantly introduced the notion of multiple causation in disease, what has unfortunately attracted most attention is the analytical problems posed by the intricate concatenations within the web. The underlying framework of the web remains biomedical individualism, or the biomedical model, that emphasises biological determinants of disease amenable to intervention through the healthcare system.

In its place Krieger first offers the social production of disease model, which 'alludes to understanding that patterns of health and disease *among* persons in these groups requires viewing these patterns as the consequence of the social relationships *between* the specified groups, with these relationships expressed through people's everyday living and working conditions, including daily interactions with others' [33]. But Krieger is concerned that the model excludes human history and origins, thereby discouraging epidemiologists from considering why population patterns of health and disease exist and persist or change over time. Furthermore, epidemiological research embodies a particular way of seeing as well as knowing the world, and this is not touched upon in the social production of disease model.

So, instead, Krieger puts forward an 'ecosocial framework' with its image of the continually constructed scaffolding of society. Here different population epidemiological profiles reflect interlinked and diverse patterns of exposure and susceptibility created by the dynamic connectedness of human existence. She presents a model specifying questions about social structure, cultural norms, ecological milieu, politics, economics and biology, and directs epidemiologists to think about individuals in the context of their everyday lives, as shaped by their intertwined histories as members of a particular society, and as biological creatures who grow, develop, interact and age.

The ecosocial framework has an evolutionary and sociological dimension, and focuses on broader determinants of health that can only be changed through more widespread social action. Finally, it encourages epidemiologists to think about themselves and their work as part of a system, not separated from it [34]:

And at the same time, by situating social groups and individuals – which, by definition, include epidemiologists – in the context of particular societies at particular times, it demands that epidemiologists consider how their social position affects the knowledge they desire and that which they produce. As such, it directs attention not only to the social production of disease, but also to the social production of science – that is, how a society's predominant view of the world and the position of scientists in this society influence the theories they develop, the research questions they ask, the data they collect, the analytic methods they employ, and the ways they interpret and report their results.

Krieger's framework may be, as she admits herself, far from fully formed but it does indicate a way forward for epidemiology and public health.

The following chapter takes the discussion about the context of QRA further, by looking at the notion of evidence, and also at philosophical and historical developments in public health.

NOTES

1. This successful book, co-written by the present Chief Medical Officer for the UK (Professor Liam Donaldson), has been considered the key textbook for those training in public health for over a decade; the title was recently changed to *Essential Public Health*.

2. The Henle–Koch postulates were criteria for assessing the causal relationship of organism to lesion. Robert Koch confirmed conditions put forward some years earlier by Jacob Henle, and acceptance of these marked the beginnings of medical bacteriology.

3. Although these terms all refer to the same methodological practice, semantically they can be separated: risk-factor epidemiology specifies the notion of risk; analytical epidemiology is differentiated from descriptive epidemiology; and modern epidemiology is the most recent and also reflects that contemporary epidemiology *is* this methodological endeavour.

REFERENCES

1. C. T. Onions, ed., *Shorter Oxford English Dictionary on Historical Principles, Vol. I*, 3rd edn. (Oxford: Clarendon Press, 1977), p. 669.

2. C. H. Hennekens and J. E. Buring, *Epidemiology in Medicine* (Boston: Little, Brown & Co., 1987), p. 3.

3. R. J. Donaldson and L. J. Donaldson, *Essential Public Health Medicine* (London: Kluwer, 1994), p. 35.

4. M. Susser and E. Susser, Choosing a future for epidemiology. I. Eras and paradigms. *Am. J. Pub. Health*, **86** (1996), 668–73.

5. M. Susser, Epidemiology in the United States after World War II: the evolution of technique. *Epidemiol. Rev.*, **7** (1985), 147–77.

6. N. Krieger, Epidemiology and the web of causation: has anyone seen the spider? *Soc. Sci. Med.*, **39**:7 (1994), 887–903.

7. D. Porter, The decline of social medicine in Britain in the 1960s. In D. Porter, ed., *Social Medicine and Medical Sociology in the Twentieth Century* (Amsterdam: Rodopi, 1997), pp. 97–120.

8. G. Taubes, Epidemiology faces its limits. *Science*, **269** (1995), 164–9.

9. D. A. Savitz, In defence of black box epidemiology. *Epidemiology*, **5**:5 (1994), 550–2.

10. R. Doll and R, Peto, *The Causes of Cancer: Quantitative estimates of avoidable risks of cancer in the United States today* (New York: Oxford University Press, 1981).

11. J. P. Vandenbroucke, Is 'The causes of cancer' a miasma theory for the end of the twentieth century? *Int. J. Epidemiol.*, **17**:4 (1988), 708–9.

12. A. M. Lilienfeld and D. E. Lilienfeld, Epidemiology and the public health movement. *J. Pub. Health Policy*, June (1982), 140–9.

13. M. Susser, Epidemiology: 'a thought-tormented world'. *Int. J. Epidemiol.*, **18**:3 (1989), 481.

14. P. Skrabanek, The emptiness of the black box. *Epidemiology*, **5**:5 (1994), 553–5.

15. G. Rose, Sick individuals or sick populations. *Int. J. Epidemiol.*, **14** (1985), 32–9.

16. N. Krieger and S. Ziegler, What explains the public's health? A call for epidemiologic theory. *Epidemiology*, **7**:1 (1996), 107–9.

17. D. Loomis and S. Wing, Is molecular epidemiology a germ theory for the end of the twentieth century? *Int. J. Epidemiol.*, **19**:1 (1990), 1–3.

18. N. Pearce, Traditional epidemiology, modern epidemiology, and public health. *Am. J. Pub. Health*, **86**:5 (1996), 682.

19. S. Wing, Limits of epidemiology. *Med. Global Survival*, **1**:2 (1994), 77.

20. *Ibid.*, 83.

21. N. Pearce, Traditional epidemiology, modern epidemiology, and public health. *Am. J. Pub. Health*, **86**:5 (1996), 679.

22. M. Susser, Judgement and causal inference: criteria in epidemiologic studies. *Am. J. Epidemiol.*, **105**:1 (1977), 1–15.

23. D. Weed, On the logic of causal inference. *Am. J. Epidemiol.*, **123**:6 (1986), 965–79.

24. A. Renton, Epidemiology and causation: a realist view. *J. Epidemiol. Comm. Health*, **48** (1994), 79–85.

25. C. Buck, Popper's philosophy for epidemiologists. *Int. J. Epidemiol.*, **4**:3 (1975), 159–68.

26. M. Maclure, Popperian refutation in epidemiology. *Am. J. Epidemiol.*, **121**:3 (1985), 343–50.

27. D. L. Weed, An epidemiological application of Popper's method. *J. Epidemiol. Pop. Health*, **39** (1985), 277–85.

28. N. Pearce and D. Crawford-Brown, Critical discussion in epidemiology: problems with the Popperian approach. *J. Clin. Epidemiol.*, **42**:3 (1989), 177–84.

29. C. Buck, Problems with the Popperian approach: a response to Pearce and Crawford-Brown. *J. Clin. Epidemiol.*, **42**:3 (1989), 185–7.

30. A. J. McMichael, Global environment change and human population health: a conceptual and scientific challenge for public health. *Int. J. Epidemiol.*, **22**:1 (1993), 1–8.

31. M. Susser and E. Susser, Choosing a future for epidemiology. II. From black box to Chinese boxes and eco-epidemiology. *Am. J. Pub. Health*, **86** (1996), 674–7.

32. D. Rapport, Changing currents in science. *Ecosystem Health*, **3**:2 (1997), 1–2.

33. N. Krieger, Epidemiology and the web of causation: has anyone seen the spider? *Soc. Sci. Med.*, **39**:7 (1994), 894.

34. *Ibid.*, 898.

8

Public health, philosophy and the nature of evidence

The time-series epidemiological studies, which underpin the quantitative risk assessment (QRA) presented in Chapter 6, aim to increase knowledge about the relationship between an atmospheric pollutant and an aspect of health or health service use. In the previous chapter developments in epidemiology which form the backdrop to time-series studies were explored. The discussion is extended in this chapter by looking at the overarching objective of such studies, to acquire evidence for or against the hypothesis of any given study, for instance that an increase in the level of a pollutant is associated with increase in an adverse health outcome or greater health service use. But what is evidence, what constitutes good evidence, and why?

This chapter first draws out some of the basic philosophical assumptions underlying the idea of scientific evidence, and probes the escalating interest and status given in Western countries to evidence-based medicine and evidence-based policy. The meaning and limitations of the nature of evidence in these contexts are examined, along with ethical pitfalls – relevance of these ideas to the QRA is explored. The second half of the chapter then brings in developments in public health practice, and illustrates how public health and environmental concerns have diverged in modern times. The implications for work such as the QRA are discussed.

Some of the issues examined in this chapter are closely connected to elements of the previous chapter, and links are highlighted.

Evidence-based medicine and evidence-based policy

What is evidence-based medicine?

Evidence-based medicine has become a burgeoning field over the last decade, and is now flavour of the month with workshops and journals as well as teaching and research programmes devoted to its principles and practice [1, 2]. But despite the exposure evidence-based medicine has received, there is still uncertainty about what it really means.

In 1991, the practice of evidence-based medicine was described as involving the following five-stage process:

1. Formulate for a chosen clinical problem an answerable question, for instance about patient diagnosis, prognosis, therapy or the organisation of services.
2. Search the medical literature and other sources for information pertaining to that question, i.e. find all available evidence.
3. Critically appraise the validity and usefulness of the evidence identified.
4. Apply the results by managing the patient accordingly.
5. Evaluate that practice [3, 4].

As straightforward as this process may sound, five years later an editorial in the *British Medical Journal* attempted to clear up some continuing confusion. First authored by a chief proponent, David Sackett, evidence-based medicine was defined as 'the conscientious, explicit and judicious use of current "best" evidence in making decisions about the care of individual patients' [5].

Both the five-stage process and the definition predicate that evidence-based medicine is an attempt, or movement toward, trying to ensure that clinical practice is based on the best available evidence. Rooted in clinical epidemiology, evidence-based medicine is supported by the recognition that a lot of contemporary medical practice (and health policy) is not based on reliable evidence. This is an important leap, because it is now over 30 years since Cochrane pointed out that the effectiveness of many medical interventions had not been properly evaluated [6].

At that time Cochrane's opinions were considered highly controversial, especially those concerning the emotive subject of cervical cytology screening for cancer of the cervix. To a lot of people the idea that medical practice and health policy might be founded on well-established errors was both difficult to believe and highly threatening. But since Cochrane's seminal ideas there has been a slowly expanding volume of work illustrating that many healthcare interventions are not based on evidence of proven effectiveness. Figures from the 1980s suggested that only 15% of medical practice was based on sound evidence, while a more recent evaluation indicated, a little more favourably, that 21% of medical technologies were evidence-based [7].

Evidence-based medicine has placed Cochrane's agenda firmly in the general medical and public domain. It acknowledges that lack of professional consensus about evidence of effectiveness exists for many clinical scenarios and that extensive variation in clinical practice occurs. However, though patients benefit from proven interventions – and are less likely to be subjected to treatments whose benefit has not been properly evaluated – the driving force behind evidence-based medicine has undoubtedly been economic.

Now well into Relman's 'era of accountability' [8], policy-makers want to know what actually works, and evidence-based medicine provides a framework necessary to limit spending and enhance efficiency. This is because evidence-based medicine is not merely the encouragement of clinicians to practice from what is scientifically known, nor just the advocacy of research in certain areas such as new technologies: evidence-based medicine is also a political endeavour.

As well as the journals, workshops and research programmes, the UK Government has set up and funded centres of excellence in Reviews and Dissemination at York and Health Technology Assessment at Southampton, and centres for evidence-based pharmacotherapy, nursing, medicine, mental health and dentistry. Politicians have placed clinical effectiveness and the related clinical governance at the forefront of current health policy [9–11]. In many health authorities, health boards and Primary Care Trusts (PCTs) in the UK, there are specialists dedicated to promoting – in the clinicians and healthcare workers serving the local population – the principles and practice of evidence-based medicine. This ethos is reflected in other Western countries.

Limitations of evidence-based medicine

While better use of evidence in medicine may be meritorious, and is undeniably appealing, the current centrality of evidence-based medicine within healthcare policy and practice needs critical reflection. In particular, and with the earlier outline in mind, it is worth looking at some of the philosophical and ethical limitations to evidence-based medicine. The aim is not to describe each in detail, as most are well rehearsed, but to illustrate the implications and constraints posed by each for air pollution epidemiology.

Hierarchy of evidence

Within evidence-based medicine there exists an ordering of scientific research studies. To satisfy the advocates of evidence-based medicine a healthcare intervention needs substantial research proof of its effectiveness, normally repeated in a variety of settings and conditions.

The research study design at the top of the hierarchy is the randomised controlled trial (RCT), perceived as the design most likely to deliver objective, value-free evidence through its random allocation of patients to treatment (or placebo) groups. This allocation process – to which the patient and research are preferably 'blinded' – minimises bias and confounding, ensuring that the research results are most likely to be due to the effect of the intervention itself, rather than extraneous factors. The RCT has been labelled the 'gold standard' of research studies.

Next in the hierarchy comes observational research studies – cohort studies[1] or case-control studies[2] – then cross-sectional (or prevalence) studies, and lastly

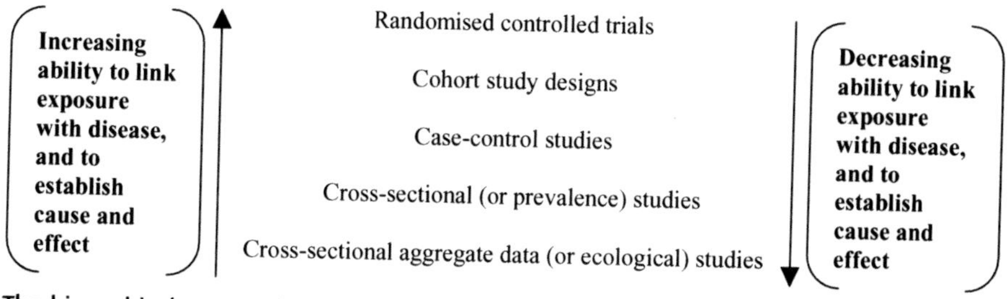

Figure 8.1   The hierarchical nature of epidemiological studies ranked by ability to establish cause and effect.
Source: C. L. Soskolne , L. E. Sieswerda and H. Morgan Scott. Epidemiologic methods for assessing the health impact of diminishing ecological integrity. In D. Pimentel, L. Westra and R. F. Noss, eds., *Ecological Integrity: Integrating environment, conservation, and health* (Washington: Island Press, 2000), p. 264.

cross-sectional aggregate data studies. Variations on this hierarchy exist, some lists adding qualitative research or peer opinion towards the bottom of the list. For instance, the US Preventive Services Taskforce rates the value of evidence from RCTs as 'grade I', non-RCTs as 'grade II' and evidence from opinions of respected authorities as 'grade III' [12]. The idea, however, remains the same, and, despite epistemological challenges [13], most orderings are basically variations on the hierarchy illustrated in Figure 8.1.

For air pollution epidemiology, the most striking aspect of this list is the place occupied by time-series studies close to the bottom of the hierarchy, meaning that they fall into the category of studies valued poorly by scientific consensus. Time-series studies are a particular kind of cross-sectional study, so named because data are collected repeatedly over a specified period. But, crucially, the data are aggregated. It is data collected at the (aggregated) population level, and *not* the individual level. So, for example, in the time-series studies discussed in Chapter 6, air pollution exposure levels are those measured in a certain region, and there is a level of presumption that individuals are actually exposed to those levels. In reality, of course, some individuals may be working outdoors by the road while others stay at home.

Statistical methods can help by adjusting, for example, for population density and mobility patterns but the fact remains that atmospheric pollutant levels are a substitute for more accurate measures of individual exposure. More specific measures have been obtained using equipment attached to individuals but studies using these are rare, expensive and time-consuming. Even using such equipment, individual exposure may still not reflect biological impact, a problem which may only be resolved in the future by measuring, say, blood levels of the pollutants.

The same issue exists for the outcome measures. Mortality and hospital admission rates are aggregated at the population level. This means that time-series studies correlate population level exposures with population level outcomes, with no way of being sure that those individuals actually exposed to high pollutant levels are those either dying or being admitted to hospital. These studies have also been called ecological studies,[3] supposedly to reflect their population level and environmental bent. This is an inappropriate label since the studies share little with ecology [14]. But it is also an unfortunate label as it has associated ecology with 'ecological fallacy', a phrase now commonly, and pejoratively, used to indicate the dangers of drawing inappropriate conclusions about individuals from population-based studies [15]:

In fact, they may be among the most difficult study designs from which to obtain valid results pertaining to individual level risks for disease. Mainly this is because we cannot be sure that the individuals who are exposed are the same ones who contract the disease. The cross-sectional aggregate data reflect only an aggregated level of exposure and disease for the group (e.g. averages), whereas a very different relationship may exist between individual exposure and disease. For this reason, we cannot necessarily extrapolate results of cross-sectional aggregate data studies directly to individual risks (i.e., the ecologic fallacy).

It is because cross-sectional aggregated data studies – such as time-series analyses – do not measure exposures and outcomes on individuals that they can be considered to be studies of limited scientific value, and so placed toward the bottom of the hierarchy. The results they provide, the evidence they yield, is open to the toughest criticism, and thereby to rejection by policy-makers.

It is here that an important distinction must be made. Rejection because of scientific methodological concerns is one thing but difficulties in accepting the evidence from time-series studies are shaped by the very nature of contemporary medicine and medical research. As discussed in the previous chapter, the dominant model in Western medicine and epidemiology remains biomedical, centred on the individual, focused on risk factors close to the individual, and geared towards individual level interventions. Research study designs have been determined by this thinking, those favoured being those orientated around individuals and apparent objectivity. The current drive towards evidence-based medicine and clinical effectiveness firmly reinforces the biomedical model, promotes simple therapeutic interventions over more complex social/behavioural ones, perpetuates the valuing of the individual above communities and, in so doing, diminishes the importance of the natural environment and human connectedness with it [16].

The epidemiological research obsession with these downstream proximate factors – those most likely to link exposure with disease and to establish cause and effect (see Fig. 8.1) – has been coined 'prisoners of the proximate' [17]. While epidemiology stays confined to this cell, time-series and other similar environmental health-related studies will remain of lower significance because, in part, they do not fit into the dominant research paradigm. The challenge is to break free from the shackles, as captured by a chapter in a recent book, *Ecological Integrity* [18], which addresses the usefulness of epidemiology to determine the links between human health, environmental concerns and ecological integrity:

In contrast to this seemingly despondent view of the place of aggregate-type studies in the world currently driven by individual risk factor epidemiology, others have begun searching for a more holistic role for epidemiology to truly address public health concerns.

The positivist nature of evidence-based medicine

Defining firm evidence and deciding what constitutes firm evidence is not a straight-forward matter, and the narrow paradigm within which scientific medicine operates is responsible for some of the conceptual confusion that lies at the heart of the effectiveness debate. The problem begins with trying to agree on what is meant exactly by firm or reliable evidence.

Most scientists would probably accept a definition which included the idea of 'weight of objective proof, about which consensus has been reached in the scientific community'. However, to some philosophers and sociologists of science, such a statement presents distinct problems, and it will be helpful here to recap some of the philosophical dimensions of scientific enquiry [19].

In the earlier parts of the twentieth century the logical positivists argued that the development of knowledge is essentially the accumulation of meaningful statements about the world (either logical propositions or statements based on empirical observation) and that progress in science is dependent on the increasing accumulation of such statements and the development of theories based on them [20].

By the 1930s, in a seminal work, Ludwig Fleck argued that every scientific concept and theory (including his own ideas on the development of the Wassermann reaction as a test for syphilis) is culturally conditioned. Fleck advocated a sociological approach to epistemology, focusing on the nature of scientific inquiry itself, rather than its logical structure. He argued that scientific disciplines develop in stages, and that long-accepted sets of beliefs (later termed paradigms) are overturned only after intellectual and practical crisis [21]. In the mid-1960s Thomas Kuhn found further evidence of paradigms in science. The existence of such long transitional stages led Kuhn to question the notion that universal standards of rationality guide scientists in their investigations [22].

Today it is widely (though not universally) accepted that we should not assume the existence of trans-historical or cross-cultural standards by which we can judge the value of scientific knowledge. In other words, we should not assume that scientific inquiry is attainably objective. If this is the case then obtaining firm evidence, if defined by scientists as necessarily objective, may not be possible. In *Contingency,*

Irony and Solidarity the philosopher Richard Rorty argues that science is just one language among many [23], each with different ways of describing the world:

... since truth is a property of sentences, since sentences are dependent for their existence upon vocabularies, and since vocabularies are made by human beings, so are truths ...

Rorty is an epistemological relativist (the idea that knowledge claims are relative to their conceptual framework), and for him the notion of an undisputed fact is problematic, as he continues:

... the world does not provide us with any criterion of choice between alternative metaphors, that we can only compare languages or metaphors with one another, not with something beyond language called 'fact'.

Many scientists are dismissive of a theory of knowledge which presents the scientific paradigm in this manner. This is unfortunate as even a basic appreciation of the argument presented might foster an understanding of, at least, the value-laden nature of scientific enquiry, and what this engenders [24]. Of particular relevance to this chapter is that if one accepts difficulties with the existence of objective fact, then a simple definition of firm evidence is equally problematic, and the same problems naturally extend to evidence-based medicine and clinical effectiveness. This explains, to a degree, what Vineis calls the irreducible fuzziness within evidence-based medicine [25].

The corollary is that research methods that see the world differently and search for different kinds of evidence are denigrated rather than being understood as alternative metaphors. Qualitative research in particular, but also theoretic evidence, expert evidence and ethics-based evidence are either found at the bottom of the hierarchy or not considered to be part of the hierarchy at all [26, 27]. Further, the awareness that changing clinical practice may be based more on beliefs than scientific evidence, and therefore requires different approaches – educational, organisational, epidemiological, behavioural, marketing, social interaction and coercive – may be sidelined or lost [28].

For patients, their values and opinions are largely removed from the equation, despite a recent upsurge in interest in clients' views of health services. By measuring what is most readily measurable, the gold standard of evidence-based medicine, RCTs, 'reduce the multi-dimensional doctor–patient encounter to a bald dichotomy ('the management of this case was/was not evidence based') and may thereby distort rather than summarise the doctor's overall performance' [29]. An alternative approach, patient-centred medicine, is humanistic, combining ethical values of the physician with psychotherapeutic theories to facilitate patients' disclosure of true concerns. It puts a strong emphasis on the patient's needs, preferences and participation in clinical decision-making yet it finds itself, as described by Bensing, in a 'separate world' to evidence-based medicine [30].

So, the effect for air pollution epidemiology is something akin to a 'double whammy'. Within the hierarchy of scientific evidence it finds itself low down, and its position is further compromised by the impact of evidence-based medicine on entrenching the perspective that the foundations of scientific medical practice are attainably objective.

Air pollution epidemiology as middle ground

Bringing together some of the different strands so far debated, it could be argued that air pollution epidemiology represents a sort of contemporary theoretical middle ground. As described in the previous chapter, historical developments in the latter half of the twentieth century reveal a rising preoccupation with technological and methodological precision in epidemiology, alongside an escalating search for associations between exposures and health parameters. Endeavours have remained largely underpinned by a biomedical model of (individual) health, focusing more on proximate risk factors, and less on what goes on inside the black box compared with associations found outside it.

However, time-series studies in air pollution epidemiology do, in some respects, represent a challenge to prevailing views. They embrace a commitment to more distal, or upstream, environmental determinants of well-being and they investigate populations rather than individuals, often over periods of sufficient length to capture a more complete picture of community experience.

Further, in the context of the lofty status afforded to evidence-based medicine, air pollution epidemiology has managed to hold its own despite the apparent lowly status of aggregated data studies. Despite the evidence provided by such studies being positioned near the bottom of the hierarchy of scientific importance, policy-makers and the public do, at times, appear to take the findings of such investigations seriously. Although not providing as dramatic an alternative to scientism as is offered by Rorty's relativism, air pollution epidemiology does begin to ask questions about the relative validity of scientific truths.

Yet such change has been countered by containment within certain boundaries. In order to be taken (at least moderately) seriously by the scientific community, air pollution epidemiology has needed to stress methodological rigour, for instance through an emphasis on control for apparent confounding factors. This detaches the population under study from its social and historical context, so placing air pollution epidemiology back into the scientific paradigm that it partly challenges. The focus on smaller and smaller pollutants within time-series studies has shown commitment within air pollution epidemiology to a reductionistic model of health, and the positivist scientific enterprise of which this is part.

So air pollution epidemiology might be seen as an example of a middle way, challenging certain tenets of the dominant scientific paradigm, yet embracing the enterprise too. The difficulty of this position is explored further by looking at evidence-based (public health) policy, and the ethical foundations of this in the positivist doctrine of utilitarianism.

Evidence-based policy: ethical and political issues

In philosophical terms evidence-based medicine is grounded in utilitarianism,[4] a particular form of consequentialism. The moral defence of evidence-based medicine hinges around the benefits of its consequences, to patients and society. Evidence-based medicine purports to identify the most effective healthcare interventions – and those least helpful or indeed harmful – which should both enable patients to make better-informed decisions, and should also be of social utility through improved efficiency and best use of scarce resources.

Against this, it is unclear exactly whose interests are included in the utilitarian calculus and how they are determined, and there is uncertainty about how to value and compare various interests. Different kinds of evidence are largely excluded, and the views of a variety of stakeholders – especially patients – may be marginalised. In addition, consequentialism may lead to conclusions held to be unethical using other moral theories, and there is no agreement over how such differences can be resolved. For instance, a utilitarian argument may conflict with a duty-based (deontological) ethical theory, the former indicating an action to be morally right when benefits are maximised, and the latter when in accord with certain duties.

For evidence-based policy-making the problems appear to be compounded. Utilitarianism as the moral theory underpinning public health policy suffers from three particular constraints. First, in calculating the benefits of the action or rule (the policy), short-term consequences tend to be more heavily weighted than those whose impact is long-term. Second, consequences closer to hand geographically are considered more relevant than those effects that are further away. And last, the consequences to human health are deemed more important than those to other sentient beings or, say, ecological integrity. All three of these have special implications for environmental health policy, where both health and environmental impacts are often longer-term and at a distance [31].

The moral dimensions of these three issues will be explored more in Part IV but one aspect of consequentialism needs to be looked at here, and that is quantifiability. There is an apparent plausibility in the assertion of proponents of evidence-based medicine and evidence-based policy that what is worthy of consideration is that which can be readily quantified. Usually these are human health effects in a short time-frame and within a discrete local or national population.

There is, however, a strong political component to the notion of quantifiability, and that relates to the issues of what *can* be readily quantified, and what policy-makers actually *want* to have quantified information on. At their core both evidence-based medicine and evidence-based policy are primarily directed at cost containment, but in the guise of information on effectiveness [32]. The desire to control expenditure is, of course, only natural, especially in the area of healthcare provision, where costs escalate with ever-increasing demand and expensive new technologies. But the values embedded within the politically and economically driven search for evidence should not be masked by the apparent objectivity of information on effectiveness. As Kerridge states [33], the concept of evidence underpinning allocation of resources may be seductive, and the desire for simplicity understandable, but only certain kinds of interventions are amenable to RCTs, the gold standard of research:

> Allocating resources on the basis of evidence may therefore involve implicit value judgements, and it may only be a short step from the notion that a therapy is 'without substantial evidence' to it being thought to be 'without substantial value'.

Certain areas of medicine are suited to RCTs, mainly the specialities, and especially those with high costs, often from new technologies. Activities 'likely to receive serious research attention are ones that result in large unit costs, with substantial short-term effect, and a limited number of well-defined alternatives' [34]. Meanwhile, general practice and primary care are less amenable because of points raised earlier; and palliative care and health care of the elderly are areas where research is traditionally difficult to do for unrelated reasons including, for instance, problems obtaining informed consent [35]. Evidence-based medicine may also impact unfairly on disadvantaged groups [36].

So political commitments are evident in evidence-based medicine, through the areas in which specific kinds of evidence can be carried out, and in the values implicit in prioritising apparent objectivity and 'de-prioritising' alternative conceptual frameworks. As Dickenson and Ashcroft postulate [37]:

> . . . it should also be noted that the state has a great interest in the success of the EBM [evidence-based medicine] programme, as a mechanism for providing an objective basis for cost control, and for defining standards of practice, which on the one hand protect the state and its employees against negligence actions, but on the other hand provide a scientific (rather than a policy or authority) based method for keeping clinicians to a clear, planned line.

Rudolf Klein is even more forthright in condemning evidence-based medicine as 'an attempt to assert the hegemony of a particular type of evidence'. He argues that it favours certain types of evidence [science], notably RCTs, over other kinds of knowledge or understanding, and esteems certain skills over others (e.g. statistics, epidemiology). Evidence-based medicine promises a 'spring-clean of existing clinical practice', introduces a brake on new interventions and technologies and offers the mirage 'of solving all health care funding problems by eliminating unnecessary, unproven and ineffective care' [38].

Evidence-based policy is modelled on evidence-based medicine, and is also keenly advocated by the Government in the UK. One aspect of the problem at the policy level is, however, that the policy-making process is far more complex than simply being determined by the available evidence. National health promotion policies in the 1990s, for example, particularly those in primary care, have been tenuously informed by evidence and determined mostly by political priorities [39]. For Klein, evidence-based policy is highly contestable and misguided, in part from the inadequacy, ambiguity, and inappropriateness of much research evidence, but also from an exaggerated claim of what science can deliver. Most significantly, however, is the 'gross misunderstanding of the policy-making process', which he describes as closer to Aristotelian practical wisdom, than any simple notion based on available evidence [38].

While evidence-based medicine emphasises the place of research evidence in clinical practice, it is only a short step – or extension – to seeing how the same factors, problems and conceptual flaws apply to evidence-based policy-making. The most obvious example may be allocation of healthcare resources, but the extension has more subtle and deep-rooted effects. These effects are most stark for health promotion policies, fiscal policies, and – most relevantly for this chapter – environmental and transport policies.

Research in these areas is hard to carry out, often not scientifically merited, politically difficult and frequently fails to inform policy, let alone result in policy change. A parallel illustration was provided by Acheson's 1998 enquiry into inequalities in health, a subject looked at further in Part IV. Alongside the enquiry itself, a high-profile group looked at the many submissions to the enquiry and found that, while there was much research demonstrating that inequalities exist, there was precious little evidence relating to the impact of specific interventions *on* health inequalities. Their conclusions [40] are relevant to this chapter:

Systematic reviews and well designed interventions have been more common in the evaluation of clinical interventions than in evaluations of social, economic, or educational policy. The fact that there is more evidence available about interventions aimed at individuals does not mean that interventions aimed at whole communities are not effective but rather reflects the paucity of good quality studies of these more 'upstream' interventions.

Clearly the sort of evidence gathered on the benefits of interventions aimed at individuals may not help guide policies directed towards reducing health inequalities. Substantial improvements in health inequalities are likely to come from reducing inequalities in wealth through progressive taxation and income redistribution.

Focusing on individual level determinants of health while ignoring more important macro-level determinants 'is tantamount to obtaining the right answer to the
wrong question' [41].

The same goes for environmental and transport policies. Substantial improvements to air pollution levels and environmental quality will likely only come about
from dramatic shifts in the way evidence is perceived and how the environment is
valued, through changed political priorities and, perhaps most fundamentally, from
radical policies. Meanwhile, the report from the *Committee on the Medical Effects of
Air Pollution* attributing almost 25 000 deaths in the UK a year to air pollution can
remain largely ignored, the *National Air Quality Strategy* can get away with paying
lip-service to the environment, and the local QRA described earlier is likely to have
little effect more than feeding rather shallow concern for the environment.

Developments in public health in the UK: separation from the environment

The historical and philosophical developments that have been discussed so far in
Part III – in environmental policy, epidemiological theory, and evidence-based
medicine and policy – bear important relation to developments in the practice of
public health. As an illustration, developments in public health in the UK are looked
at here.

Public health in the UK has had a unique history. Born in the midst of urbanisation and industrialisation, public health became rapidly dominated by the medical
profession, and has gone through – and continues to experience – periods of decline
and resurgence. There have been numerous changes of name, frequent identity
crises and, particularly through the twentieth and twenty-first centuries, ongoing
insecurities. Of most relevance to this chapter, however, is that public health has
gradually – and on occasions suddenly – been separated from its theoretical and
practical relationship with the environment. For work on air pollution today this
means that public health has become somewhat restricted, in terms of how it can
be involved and what it can achieve.

Public health in proximity to environmental issues

As discussed in Part I, the huge influx of people to cities in the eighteenth and
nineteenth centuries, accompanied by changing working practices and conditions,
brought new and worrying health problems. Infectious diseases spread easily in
areas where people lived in close proximity and dreadful sanitary conditions, and
the environment of factories and industries heralded occupational diseases and
accidents on an unprecedented scale, and special threats to child health and development. Vaccination against smallpox, which was very common at that time, was

the only statutory measure enforced upon local authorities, but cholera epidemics were a similarly huge concern.

These anxieties, accompanied by the need to have a healthy workforce in an era of an expanding British Empire, engendered a search for possible solutions. Two linked debates informed this search. The first was about infectious disease causation, whether miasma or contagion was aetiologically responsible. This debate was fierce but theoretically (and from the position of participants) not as dichotomised as has sometimes been presented. The second debate was about morality, in particular the moral nature of the poor. Darwinian ideas were being extended to the social domain substantiating the idea, among some, that those less fit be selected out in the interests of the moral advance of the nation. This supported allowing those morally inferior – such as the poor – to either languish or be weeded out through eugenic policies, and also justified aggressive imperialist action towards other nations and cultures. These debates were covered in more detail in Part I.

Whatever the reasons for the behaviour of the poor towards hygiene, infectious diseases did not respect boundaries and the workforce needed to stay healthy. Believing that health improvement lay beyond the scope of the medical profession, pioneering lawyers and engineers, such as Edwin Chadwick, advocated sanitary reform as the key in the mid-nineteenth century. Despite coming up against influential individuals and companies (such as the water authorities), these innovators were not promoting ideas in conflict with political will, but rather the opposite. This, as will be returned to later, contrasts with the more contemporary position of public health practitioners taking on environmental issues.

Removal of environmental filth and improvements to sanitation would check the progress of infectious diseases, but the Government soon felt that the medical profession should both determine how this could be achieved, as well as monitor the nation's health. The position of Medical Officer of Health (MOH) was created to advise and monitor locally, and John Simon (who had been one of the initial metropolitan MOHs) became the first adviser to, and planner for, central government. As the appointment of MOHs became mandatory – to metropolitan sanitary districts in 1855 and to provincial districts in England and Wales in 1875 – a State medical service had essentially been established. It was dominated and controlled by the medical profession and some would argue has remained so in practice to this present day [42].

As the last quarter of the nineteenth century progressed, although no speciality existed as such, public health consolidated itself. In 1871 a postgraduate certificate was developed, which later became the Diploma in Public Health (DPH), first in Trinity College Dublin, then Edinburgh and Cambridge. By 1876 Oxford and London had a certificate in Preventive Medicine and Public Health. The DPH, which existed for just over one hundred years, was only available to members of the medical profession and could be registered with the General Council for Medical Education and Registration.

Although DPH courses varied a little depending on university the essential components were, and remained long-term, a combination of classroom and practical work. While the former consisted of lectures and study time, later accompanied by laboratory work, the latter involved a period of experiential training alongside an existing MOH, a method mirroring that practised in clinical medicine. Environmental issues were integral to learning. Both meteorology and climatology were required parts of the initial DPH curricula, receiving great emphasis, and stayed there for some decades. Whether one was a miasmatist or contagionist the weather was felt to affect the spread of infectious disease epidemics, and William Farr placed meteorological data alongside the published annual national health statistics, which were not abolished until 1973 [43].

Those training to be, and taking up posts as MOHs, were a mixture of experienced doctors and others going into the speciality soon after completing their medical training: some worked full-time and some part-time. Around the birth of public health the position of MOH carried gravitas as well as political weight and power, and attracted high-calibre physicians hoping it would boost their private practice and status. However, although a public health specialist qualification was made compulsory for metropolitan districts in 1891, by the end of the century the status of the MOH had declined. Those taking up posts tended to be either general practitioners wanting – or needing – to supplement their private income, or those attracted by the lure of a steady, albeit unremarkable, income from local government [44]. Despite this, naturally, there were some drawn by a passion for the job.[5]

At the turn of the century MOHs were well-established positions in local authorities. Working alongside sanitary inspectors, engineers and those responsible for housing and town planning, they monitored the health of their local communities (including notification of infectious diseases from 1899), oversaw the health dimensions of local developments, were involved in industrial and occupational matters, and acted as advocates for health improvement, such as through reduction in smoke pollution. Despite significant changes to the direction and nature of public health over the coming periods, these essential components of the public health doctor remit – enabled by the location and working relations of the MOH – remained present and active until the 1970s.

As the twentieth century dawned, what did shift was the emphasis in public health towards preventive medicine. Once the germ theory of disease was largely accepted, awareness grew of the importance of personal and social habits in the spread of infectious diseases. In tandem with this, concern mounted about national efficiency and the nation's health.

In response to these developments environmentalists demanded wider preventive services. While eugenicists unsuccessfully countered these calls,[6] services were set up to advise, inform, improve and monitor the health of groups in need, especially the young who were seen as both the nation's future and also the starkest recipients of the detrimental health effects of poor hygiene. So welfare programmes for pregnant women and post-natal mothers, infants and schoolchildren were established, as well as midwives and health visitors, the latter operating as hygiene instructors in the early years rather than as a resource for new mothers (as today) [45].

These services, later alongside community health centres and social work, created teams based in public health departments, under the leadership of MOHs, which oversaw the health of the local community. But the traditional responsibilities of the MOH, and hence public health, for environmental health and as 'community watchdog' remained, as these officers were still best placed, and most suitably trained, to monitor and act accordingly. Strong allegiances existed between MOHs and, say, town planners, for instance in the creation of integrated municipal management systems to ensure health efficiency [42].

The tasks and duties placed on the MOH and his or her department at this time were numerous and wide-ranging, but it was a good time for public health. When Lloyd George introduced the National Insurance Act in 1911, the numbers able to use municipal hospitals, then under the administration of MOHs, increased, as did the status of these hospitals (in comparison with the ailing voluntary hospitals), so further adding to the responsibilities of the MOH and the good feeling within the profession. The inter-war period has been labelled by some commentators as the golden era of public health [45].

There is therefore an apparent paradox, created from the seemingly conflicting notions that public health both lost status after about 1900, and that the 1920s and 1930s were a special, and good, time for the profession. This can be reasoned as follows. The loss of status was related to the change of direction away from surveillance and towards community health: individual prevention involving mainly mothers and children. Part of the loss was that this was seen to be primarily women's work, and this resonates through the lowly position that community health still occupies today. On the other hand, however, management of the constantly expanding State medical service reached its maximum extent during the period 1929–48 when the Poor Law medical services were incorporated into public health. This made the MOHs very powerful but at the same time had negative professional connotations if public health is conceived to be principally about surveillance of health rather than management of medical care.

In 1946 the National Health Act described a new health service that would become operational two years later. Planning, however, had been going on before and during the Second World War, and MOHs had envisaged a three-tiered service with themselves in the coveted position of central coordinators and administrators. But Government plans dashed all hopes of being at the heart of the new system and, instead, public health was placed at the periphery and disempowered. Bevan, along with most politicians and doctors at the time, equated improvements in medical care with improvements in health, a fallacy unmasked a few decades later but still widely subscribed to today [46]; preventive medicine was relatively unimportant. The new service was orientated around State hospitals, which replaced existing municipal and voluntary hospitals, with teaching hospitals dominating the system. General practitioners would act as the gateway to the new establishments [47].

Hospital consultants were not unhappy. Voluntary hospitals were in demise anyway, private practice could continue, and the merit award scheme would compensate high achievers (and earners) for revenue lost through obligatory participation in the National Health Service (NHS). General practitioners also gained assurance regarding their role and position, and a reliable salary with detachment from the medical market-place was appealing to many.

Public health, however, was marginalised. It remained outside the NHS, based in local authorities, with its functions substantially reduced and its status severely compromised. For some of the community clinical services it had to compete with general practice. MOHs and their public health departments were pressured from both within and without the medical profession, and morale fell. It could be argued [48] that, despite a brief resurgence in the early 1970s [49], it is a position from which public health has never fully recovered:

Just as the public health doctor was not able successfully to justify his continued work in personal prevention, so he failed to make a good case for the medical administration of welfare work. Thus MOsH increasingly found themselves accused of failure in respect to the delivery of effective community care, and squeezed between the twin pressures of general practice from without and social work from within.

Regardless, however, of the enthusiasm or mental state of its practitioners, public health remained – and continues to remain – a vital function. Responsibility for environmental health, among other important functions, stayed with public health departments in their local authority base. MOHs still held and built empires in their departments, although the prestige of these may have declined.

But, as Jane Lewis has pointed out, public health lacked a philosophy. Apart perhaps from at its inception, public health had always defined itself by the tasks given to it, rather than by the development of a coherent theoretical and practical framework through which to debate and act. This seemed unimportant when things were going well, but when tasks were taken away a vacuum of identity, articulation and purpose was created [50].

Social medicine

In the 1950s and 1960s an opportunity arose to fill that void. Social medicine offered the possibility to both bring together service and academic workers in public health, and to provide a theoretical framework through which a unified speciality could develop. But social medicine meant different things to different people at different times. Originally conceived as a sociobiology of health and disease, early proponents of social medicine postulated a new type of doctor, who sought to understand his or her patients – their health and disease – in the context of their communities, their environment, and their personal and social histories. Medical training needed to be revamped to create a new breed of clinicians, who practised whole-person medicine, a holistic endeavour combining clinical and preventive medicine and underpinned by a framework of social ecology [51, 52].

Academics, however, moved social medicine in a different direction. They conceptualised it, not so much as scientific humanism but as a positivist intellectual enterprise to explain the social basis of illness in populations. Founded on scientific rationalism and epidemiology, social medicine would describe how health and disease in societies was produced, including social, environmental and biological determinants. Early proponents were caught in an ideological difficulty as they also largely held that medical care would solve health problems once it was available universally. But by the 1960s and 1970s this was challenged, both by McKeown's thesis that many of the health improvements in the late nineteenth and early twentieth centuries were due to improvements in sanitation and living conditions rather than the effects of medicine, and also with the introduction of qualitative methods into social medicine.

The rifts, both within social medicine itself and between its academic proponents and service-based public health practitioners, could not be reconciled. When major reorganisations of the NHS and public health were planned and implemented in the early 1970s, social medicine was sidelined. It did not, however, die and, as Dorothy Porter has highlighted [53], social medicine has had something of a revival recently with new university departments in Bristol and Birmingham, reflecting both a resurgence of interest and also the philosophical leanings of individuals.

Public health separated from environmental issues

At the end of the 1960s public health was at yet another crossroads, but this juncture was to be highly significant for environmental health. Escalating costs of healthcare delivery and new medical technologies, accompanied by a realisation that demand for healthcare services was ever-increasing, prompted the governments of many countries to think of ways to contain costs.[7] In the early 1970s the UK Government planned a major reorganisation of the NHS which eventually saw the creation of area health authorities (coterminous with local government units responsible for social services) with 14 regional health authorities responsible for planning, and 205 district management teams below. Managerial negotiations would go on between health authorities and hospitals as to the level, nature and, of course, cost, of provision of services. But who would arbitrate or broker the arrangements [47]?

Public health doctors, it was felt, might be in the best position to do this. As members of the medical profession they spoke the same language as their clinical colleagues, and enhanced training in administration, management and organisation of health services would serve them well. And a further carrot was dangled, one that would improve their standing within the profession: speciality status. Following negotiations between various public health professional groups and the General Medical Council, a new faculty was set up in 1972 under the auspices of the Royal College of Physicians – the Faculty of Community Medicine (FCM).[8]

Alongside the FCM a new medical speciality was created, that of community medicine, with a training and examination process (Membership of the Faculty of Community Medicine) paralleling the clinical specialities. This step was appealing to some public health doctors, who felt their credibility would be boosted, and a new identity formed; others, however, were resentful or took early retirement. It was anticipated, nevertheless, that community physicians would be the medical professionals of community health, monitoring the health of their local communities, assessing the need for health care, evaluating services and acting as important administrators. In fact, the honeymoon period was short-lived and within a few years community medicine was confused and demoralised. Although there is insufficient scope to discuss that further here, Lewis propounds [54]:

. . . little thought was given to the way in which community medicine would be practised in the new NHS and community physicians experienced considerable tension in reconciling first their responsibility for the management of health services with that of analysing health problems and, second, their formal accountability to the NHS bureaucracy with their ethical accountability to their communities.

What is most relevant, however, is that public health – now as community medicine – moved out of local authorities and into the NHS. Local authorities, pushed to one side in the reorganisations, were left with the vital health-related areas such as environmental health and community care. The position of MOH had been abolished, and existing public health doctors became community physicians or medical officers at a variety of levels. Occasional posts may have remained outside the NHS but essentially public health split with local government, and newly qualified community physicians would be predominantly based in public health departments in NHS health authorities [55].

With this move, public health doctors and their departments became irrevocably separated from their colleagues working on matters of environmental health, sanitation, and local planning and development. Working contacts may have been retained for a while but inevitably over time relationships disintegrated, the decline accelerated by the new workloads of public health departments as well as the different geographical boundaries within which the two authorities operated: it is difficult to coordinate activities when the district community of the health authority is defined differently to that of the corresponding local authority.

In 1982 another NHS reorganisation abolished the area tier, further polarising community medicine (now uncertain as to whether prime responsibility lay with the management of health services, or analysis of health problems and needs), although the 1988 enquiry into the state of public health by Sir Donald Acheson [56] bolstered morale within the speciality as well as resulting in a name change to public health medicine.[9] Major changes in healthcare organisation and financing under Margaret Thatcher in the 1990s fuelled more recent developments in public health, which were discussed in the Introduction of the book.

Crucially, however, the divide forged by the 1974 reorganisation has never been bridged. While those in local authorities have reinforced their training and backgrounds – as Environmental Health Officers (EHOs, previously sanitary officers), town planners and public health engineers (a subspeciality of civil engineering) – these skills have become harder for public health workers to tap in to [52]. It is only recently, with the very latest restructuring of the health service (and public health), that the prospects of collaborative working have resurfaced. This is discussed further in the conclusions of this book.

The impact on public health practice

The strategies needed to tackle environmental issues such as air pollution are necessarily cross-disciplinary, requiring the kind of multisectoral approach firmly advocated in recent high-profile UK Government documents [57–59]. In conjunction with the QRA described in Chapter 6, some of the local initiatives taken to address outdoor air pollution and health in Barking and Havering Health Authority included the establishment of an Air Quality and Health Group (a joint venture between the public health department of the health authority and the environmental health departments of the local authorities, aiming to develop local strategies to tackle air quality problems, e.g. improved cycleways and workplace-centred initiatives) [60]; the establishment of an air pollution warning system to alert the public of poor air pollution episodes [61]; the creation of an air pollution freephone telephone number; and the development and distribution of air pollution information leaflets, given that public awareness of the causes and health effects of air pollution may be poor and laden with misconceptions [62].

However, such initiatives need to be treated with caution. They are likely to continue only as long as local enthusiasm waxes, and their impact will be limited by the methodological deficiencies of the work, the underpinning problems with epidemiological theory and policies on effectiveness, and the organisational relationships that work against the operational success of these kinds of activities.

Furthermore, little improvement in air quality is likely from local action. In London, perhaps most potential lies with the relatively new mayor who – with the Greater London Authority – will be responsible for 'promoting economic and social development in London and improving the environment' [63], offering the opportunity to embrace London's particular problems of air pollution, transport and health [64, 65]. As mentioned before, early signs suggest that congestion charging[10] may be impacting significantly on traffic levels in central London.

Most important, however, are national policies, which shape local and regional action, and need to be more committed to protection of population health. Unfortunately, the *National Air Quality Strategy* set air quality standards and objectives in 1997 which were weak and poorly enforced at the time. This suggests that the Government considered the existing levels of morbidity and mortality associated with air pollution to be acceptable.

NOTES

1. A cohort study follows a group, or cohort, of individuals and compares the incidence of disease in those members of the cohort exposed to a risk factor with the incidence in those unexposed. Cohort studies are usually time and resource intensive. They are observational since no specific intervention is applied.
2. A case-control study finds individuals with a disease (the cases) and compares their exposure to a risk factor with a comparison group of individuals who do not have the disease (the controls). Case-control studies are usually quicker and cheaper than cohort studies, and are also observational.
3. They are also sometimes referred to as correlational studies as they correlate aggregated data at the population level.
4. Utilitarianism is discussed in more detail in Chapter 9.
5. General practitioners operated on a fee-for-service basis, and there was competition for patients and income. MOHs were salaried by local government although, in time, salaries were found to vary between locations, sometimes dramatically. Security was not such a draw, as tenure of appointment did not occur for some time.
6. Some preventive services could be supported by eugenicists, e.g. family planning.
7. When the NHS was set up, many believed that the need for health care could be fairly met by an improved system, and with efficiency gains. In the following decades, however, there was a growing understanding of the difference between demand and need for health care, and an appreciation that there were always opportunities for additional health gains, albeit at perpetually rising marginal costs.

8. There had been discussion about an umbrella faculty being multidisciplinary, housing public health workers from different backgrounds, but it was felt this would not be allowed under the Royal College of Physicians, so the 'doctors-only' option was preferred.

9. In 2003 a vote was held, results favouring a change of name from the Faculty of Public Health Medicine (FPHM) to the Faculty of Public Health (FPH).

10. Congestion charging was introduced in February 2003, applicable to most vehicles entering a 'congestion zone' in central London.

REFERENCES

1. Doctors in Britain urged to practise effectively [News]. *BMJ*, **312** (1996), 143–4.

2. Evidence based medicine: in its place [Editorial]. *Lancet*, **346** (1995), 1171–2.

3. D. L. Sackett, R. B. Haynes, G. H. Guyatt and P. Tugwell, *Clinical Epidemiology: A basic science for clinical medicine* (London: Little, Brown, 1991), p. 187.

4. North Thames Regional Health Authority Annual Public Health Report, *Partnerships for the Future: Health and healthcare* (North Thames Regional Health Authority, 1996), p. 21.

5. D. Sackett, W. Rosenberg, J. Gray *et al.*, Evidence based medicine: what it is and what it isn't [Editorial]. *BMJ*, **312** (1996), 71–2.

6. A. Cochrane, *Effectiveness and Efficiency* (London: Nuffield Provincial Hospital Trust, 1972).

7. T. Greenhalgh, Is my practice evidence based? *BMJ*, **313** (1996), 957–8.

8. A. Relman, Assessment and accountability: the third revolution in medical care. *N. Engl. J. Med.*, **319**:18 (1988), 1220–2.

9. NHS Executive, *Promoting Clinical Effectiveness: A framework for action in and through the NHS* (Leeds: NHS Executive, 1996).

10. L. J. Donaldson, Clinical governance: a statutory duty for quality improvement. *J. Epidemiol. Comm. Health*, **52** (1998), 73–4.

11. Department of Health, *Clinical Governance: Quality in the new NHS* (London: HMSO, Health Service Circular 1999/065, 1999).

12. US Preventive Services Taskforce, *Guide to Clinical Preventive Services*, 2nd edn (Baltimore: Williams & Wilkins, 1995), p. 862.

13. R. E. Ashcroft, Current epistemological challenges in evidence based medicine. *J. Med. Eth.*, **30**:2 (2004), 131–5.

14. A. J. McMichael, The health of persons, populations, and planets: epidemiology comes full circle. *Epidemiology*, **6**:6 (1995), 633–6.

15. C. L. Soskolne, L. E. Sieswerda and H. Morgan Scott, Epidemiologic methods for assessing the health impact of diminishing ecological integrity. In D. Pimentel, L. Westra and R. F. Noss, eds., *Ecological Integrity: Integrating environment, conservation, and health* (Washington: Island Press, 2000), p. 267.

16. A. Liberati and P. Vineis, Introduction to the symposium: what evidence based medicine is and what it is not. *J. Med. Eth.*, **30**:2 (2004), 120–1.

17. A. J. McMichael, *Prisoners of the proximate: loosening the constraints on epidemiology in an age of change.* Keynote paper presented to the Annual Conference of the Society for Epidemiologic Research, Chicago, 1998.

18. C. L. Soskolne, L. E. Sieswerda and H. Morgan Scott, Epidemiologic methods for assessing the health impact of diminishing ecological integrity. In D. Pimentel, L. Westra and R. F. Noss, eds., *Ecological Integrity: Integrating environment, conservation, and health* (Washington: Island Press, 2000), p. 272.

19. For an introduction to the subject see R. Harre, *The Philosophies of Science* (Oxford: Oxford University Press, 1972); A. F. Chalmers, *What Is This Thing Called Science?* (London: Open University Press, 1982), W. H. Newton-Smith, *The Rationality of Science* (London: Routledge & Kegan Paul, 1981).

20. P. Bracken, Post-empiricism and psychiatry: meaning and methodology in cross-cultural research. *Soc. Sci. Med.*, **36** (1993), 265–72.

21. L. Fleck, *Genesis and Development of a Scientific Fact* (trans. F. Bradley and T. J. Trenn) (Chicago: University of Chicago Press, 1979). [Originally published in 1935.]

22. T. S. Kuhn, *The Structure of Scientific Revolutions* (Chicago: University of Chicago Press, 1971). [First published in 1962].

23. R. Rorty, *Contingency, Irony and Solidarity* (Cambridge: Cambridge University Press, 1989).

24. M. Risjord, Relativism and the social scientific study of medicine. *J. Med. Phil.*, **18** (1993), 195–212.

25. P. Vineis, Evidence-based medicine and ethics: a practical approach. *J. Med. Eth.*, **30**:2 (2004), 126–30.

26. S. Buetowa and T. Kenealy, Evidence-based medicine: the need for a new definition. *J. Eval. Clin. Prac.*, **6**:2 (2000), 85–92.

27. R. S. Barbour, The role of qualitative research in broadening the 'evidence base' for clinical practice. *J. Eval. Clin. Prac.*, **6**:2 (2000), 85–92, 155–63.

28. R. Grol, Beliefs and evidence in changing clinical practice. *BMJ*, **315** (1997), 418–21.

29. T. Greenhalgh, Is my practice evidence based? *BMJ*, **313** (1996), 958.

30. J. Bensing, Bridging the gap: the separate worlds of evidence-based medicine and patient-centred medicine. *Patient Ed. Council*, **39**:1 (2000), 17–25.

31. P. Vineis, Environmental risks: scientific concepts and social perception. *Theoret. Med.*, **16** (1995), 1–17.

32. R. K. Lie, Research ethics and evidence based medicine. *J. Med. Eth.*, **30**:2 (2004), 122–5.

33. I. Kerridge, M. Lowe and D. Henry, Ethics and evidence based medicine. *BMJ*, **316** (1998), 1151–3.

34. L. Culpepper and T. T. Gilbert, Evidence and ethics. *Lancet*, **353** (1999), 829.

35. R. Vos, D. Willems and R. Houtepen, Coordinating the norms and values of medical research, medical practice and patient worlds – the ethics of evidence based medicine in the orphaned fields of medicine. *J. Med. Eth.*, **30**:2 (2004), 166–70.

36. W. A. Rogers, Evidence based medicine and justice: a framework for looking at the impact of EBM upon vulnerable or disadvantaged groups. *J. Med. Eth.*, **30**:2 (2004), 141–5.

37. D. Dickenson and R. Ashcroft, *Country Report for Evibase State-of-the-art Workshop* (Paper presented at Evibase state-of-the-art workshop, Maastricht (The Netherlands), 2001).

38. R. Klein, From evidence-based medicine to evidence-based policy. *J. Health Serv. Res. Pol.*, **5**:2 (2000), 65–6.

39. D. Florin, Barriers to evidence based policy. *BMJ*, **313** (1996), 894–5.

40. S. Macintyre, I. Chalmers, R. Horton and R. Smith, Using evidence to inform health policy: case study. *BMJ*, **322** (2001), 222–5.

41. G. Davey Smith, S. Ebrahim and S. Frankel, How policy informs the evidence. *BMJ*, **322** (2001), 184–5.

42. E. Fee and D. Porter, Public health, preventive medicine, and professionalization: Britain and the United States in the nineteenth century. In E. Fee and R. M. Acheson, eds., *A History of Education in Public Health: Health that mocks the doctors' rules* (Oxford: Oxford University Press, 1991), pp. 15–43.

43. R. Acheson, The British Diploma in Public Health: birth and adolescence. In E. Fee and R. M. Acheson, eds., *A History of Education in Public Health: Health that mocks the doctors' rules* (Oxford: Oxford University Press, 1991), pp. 44–82.

44. D. Porter, Stratification and its discontents: professionalization and conflict in the British public health service, 1848–1914. In E. Fee and R. M. Acheson, eds., *A History of Education in Public Health: Health that mocks the doctors' rules* (Oxford: Oxford University Press, 1991), pp. 83–113.

45. —, *Health, Civilization, and the State: A history of public health from ancient to modern times* (London: Routledge, 1999).

46. T. McKeown, *The Role of Medicine* (Oxford: Blackwell, 1979).

47. V. Berridge, *Health and Society in Britain Since 1939* (Cambridge: Cambridge University Press, 1999).

48. J. Lewis, The public's health: philosophy and practice in Britain in the twentieth century. In E. Fee and R. M. Acheson, eds., *A History of Education in Public Health: Health that mocks the doctors' rules* (Oxford: Oxford University Press, 1991), pp. 195–229.

49. —, *What Price Community Medicine? The philosophy, practice and politics of public health since 1919* (Brighton: Wheatsheaf, 1986), p. 59.

50. —, *What Price Community Medicine? The philosophy, practice and politics of public health since 1919* (Brighton: Wheatsheaf, 1986).

51. N. Oswald, Training doctors for the National Health Service: social medicine, medical education, and the GMC 1936–48. In D. Porter, ed., *Social Medicine and Medical Sociology in the Twentieth Century* (Amsterdam: Rodopi, 1997), pp. 59–81.

52. R. Acheson, The British Diploma in Public Health: heyday and decline. In E. Fee and R. M. Acheson, ed., *A History of Education in Public Health: Health that mocks the doctors' rules* (Oxford: Oxford University Press, 1991), pp. 272–313.

53. D. Porter, The decline of social medicine in Britain in the 1960s. In D. Porter, ed., *Social Medicine and Medical Sociology in the Twentieth Century* (Amsterdam: Rodopi, 1997), pp. 97–120.

54. J. Lewis, *What Price Community Medicine? The philosophy, practice and politics of public health since 1919* (Brighton: Wheatsheaf, 1986), p. 162.

55. M. Jeffreys and J. Lashof, Preparation for public health practice: into the twenty-first century. In E. Fee and R. M. Acheson, eds., *A History of Education in Public Health: Health that mocks the doctors' rules* (Oxford: Oxford University Press, 1991), pp. 314–35.

56. Department of Health and Social Security, *Public Health in England: The report of the committee of enquiry into the future development of the public health function* [Chairman Sir Donald Acheson] (London: HMSO, 1988).

57. Department of Health, *The New NHS: Modern, dependable* (London: HMSO, 1997).

58. —, *Our Healthier Nation* (London: HMSO, 1998).

59. —, *Chief Medical Officer's Project to Strengthen the Public Health Function in England* (London: HMSO, 1998).

60. Annual Report of the Director of Public Health 1998/99, *Who's Responsible for Health in the Community? Air pollution* (London: Barking and Havering Health Authority, 1999), pp. 24–6.

61. A. S. Kessel and K. Padki, Air pollution episode warning system in Barking and Havering Health Authority. *Chem. Incident Rep.*, October (1999), 18–19.

62. Health Education Authority, *Air Pollution: What people think about air pollution, their health in general, and asthma in particular* (London: HEA, 1997).

63. R. Harling, London's health: a role for the new mayor. *BMJ*, **318** (1999), 478–9.

64. N. Soderland, J. Ferguson and M. McCarthy, *Transport in London and the Implications for Health* (London: Health of Londoners Project, 1996).

65. R. Ashton, Evaluating the wider impact on health and quality of life. In A. Fletcher and A. McMichael, eds., *Health at the Cross-roads: Transport policy and urban health* (London: John Wiley & Sons, 1997).

Conclusions to Part III

Parts I and II of this book charted the holistic conceptions of air and medicine in ancient civilisations, then the growing reductionism from miasma to smoke pollution in scientific medicine. In Part III of the book the dominant conception of air and health in Western medicine has been followed to the present day but with an additional dimension.

A case-study approach has been used to illustrate the contemporary conception of air as air pollution, and also to demonstrate what this reveals about current ways of thinking in public health. The QRA described in Chapter 6, and the epidemiological research designs contained within it, show that air has been further reduced to its constituent components, and polluted air is perceived as these parts, whose impact on the health of a community can be calculated.

But the QRA is deeply constrained by both its science, and the philosophical and policy dimensions that shape its science. As a research scientific process the QRA may have methodological flaws, but the epidemiological framework that determines those flaws is itself ill-conceived. The policy initiatives that promote certain notions of evidence and effectiveness are ethically and philosophically contestable, so compounding the problem.

At a practical level, the possibility of valuable local public health work on air pollution and other environmental issues is limited. Historical developments in public health, in much of the UK at least, have driven apart relationships among those working in environmental health. What we are left with is rather shallow environmentalism, theoretically problematic and practically restricted.

New Horizons

Overview to Part IV

The final conception of air and health is contemporary but also looks to the future. Up to now, the book has considered three broad conceptual stages in the history of air and health. In the first, relating predominantly to Greek medicine and philosophy, but also to other early civilisations, air formed part of understanding health as harmony between mankind and nature, which itself reflected a higher balance in the order of the cosmos.

In the second, early scientific stage, the spiritual or religious dimensions to the conception were gradually lost as air became understood as part of the aetiological jigsaw puzzle of infectious disease causation. Rival theories may have differed as to the specific role played but polluted air was the focus around which an important scientific debate about spread of disease was oriented. For city-dwellers, however, smoke-filled skies represented everyday anathema, poisonous palls damaging physical and psychological health, child development, vegetation, buildings and the economy, as well as drastically reducing sunlight hours.

As early epidemiological efforts to demonstrate links between polluted air and ill health failed to lead to change in policy, energies turned elsewhere. In Britain campaigners organised themselves into groups, held exhibitions, produced documents, articles and journals, and lobbied parliamentarians. Although not central to the campaign public health contributed energetically, both through the academic and special skills offered by its practitioners and through the professional weight that Medical Officers of Health (MOHs) brought to the anti-smoke lobby. In the early decades of the twentieth century public health, and its leaders in particular, enjoyed a heyday that cemented the importance of the role of community advocacy.

In the third historical stage – the modern scientific era of the second half of the last century – air has been gradually reduced to its constituent components, and the causal place of these in chronic diseases and their acute exacerbations. Research efforts – such as quantitative risk assessment (QRA) – attempting to demonstrate or apply associations between atmospheric components and putative health effects, exemplify both the reductionism, and also the parochialism, of present epidemiological thinking. Methodological and theoretical problems at the heart of environmental epidemiology represent the lack of a philosophical framework for public health, a deficiency augmented by historical developments in public health in the UK, and compounded by the antithetical current political drive for evidence-based medicine.

The most recent conception of air and health has emerged against this backdrop. In the approach taken to dealing with climate change, the debate about air has provided the opportunity to rethink the relationship between mankind and nature, and the moral dimensions of public health theory and practice. Often referred to synonymously as greenhouse warming, climate change presents an instance of the health effects of Western lifestyles being borne by those at a distance in time and place. Unlike, say, passive smoking, those affected by climate change may have little or no connection with the perpetrators yet are left with the consequences. This raises fundamental questions about the geographical, temporal and moral boundaries of public (health) responsibilities, as well as the place of utilitarianism in public health theory.

Yet underneath such debate it is hard not to feel that the real root of the problem lies in the progressive devaluing of nature that has accompanied Western, and ultimately global, economic development. Although there is often a vague and sometimes misinformed sentimentality to reflections on the environmentally-friendly approaches of our forebears, there was certainly a more holistic and spiritual dimension to the relationships of early civilisations with nature (and also for some First Nation populations today). The relatively new academic field – environmental ethics – argues for a reorientation back to more traditional values. Environmental ethicists have used climate change to illuminate their arguments; whether this offers something useful to public health is explored.

Using the approach taken in dealing with climate change as a representation of the most recent conception of air and health, Part IV is structured as follows. In Chapter 9 there is an overview of the science of climate change and its effects on human health. Utilitarianism is then considered as the moral foundation of public health, followed by critiques of utilitarianism, and the relevance of these critiques to climate change.

In chapter 10 John Rawls's alternative theory of social justice is put forward, followed by an exploration of how this has been used within the climate change debate in the guise of climate justice. The related illustration of inequalities in health is also examined. Other challenges to utilitarianism include the perspectives of Ludwig Wittgenstein, as well as proponents for a return to virtue ethics, both of which highlight serious concerns about the overall direction of modern Western moral philosophy.

In bringing together the arguments within Part IV, Chapter 11 captures an essence of relevance to the whole book. The concerns presented in Chapters 9 and 10 have historical roots which connect to other developments in the history of science, medicine and political philosophy. The final conception of air and health – the approach to dealing with climate change – can be seen as illustrative of these interlinked historical developments. Finally, environmental ethicists argue that the lack of substantial environmental thinking in Western cultures has roots that are intertwined with the developments mentioned above, and deep roots require radical solutions.

9

Climate change: science and utilitarianism

Climate change is one of a number of large-scale anthropogenic processes often collectively labelled global environmental change (GEC). Other processes under this umbrella include stratospheric ozone depletion, acid rain, deforestation and loss of biodiversity. Although no precise definition of GEC exists, the term encompasses detrimental environmental effects consequent to activities accompanying human development, most of which are relatively recent in terms of the history of the planet. The processes are often interlocked and inevitably have ramifications for human health.

Climate change: the scientific background

Climate change is often referred to interchangeably as greenhouse warming, the greenhouse effect or even global warming. Although each refers essentially to the same process, the greenhouse effect describes the bio-geo-atmospheric process, greenhouse warming and global warming refer to the resultant heating, and climate change refers to the consequences, which include changes in temperature and rainfall.

The greenhouse effect is a highly complex, incompletely understood, process that involves the geology and biology of the earth, as well as the oceans, the atmosphere and the sun. But the basic effect is a natural one; it is human activities that have distorted the natural effect. The scientific details need only be sketched for the purposes of this book.

Late in the nineteenth century a Swedish scientist, Svante Arrhenius, first predicted that a build-up of carbon dioxide in the atmosphere from the burning of fossil fuels would result in global warming. Further scientific debate occurred in the 1920s, and in 1937 the term 'greenhouse effect' was used 'to describe how atmospheric gases stabilise the earth's temperature by allowing the passage of visible and UV [ultraviolet] radiation from the sun, which warms the earth's surface, but block the escape to space of reflected infrared radiation' [1].

This early proposition is not that dissimilar to what is understood of the process today. The earth's atmosphere freely admits short-wave solar radiation from the sun, including visible light. Most of this incoming radiation is absorbed by the earth's surface and warms it. Some solar radiation is reflected off the earth's surface back to space, but some longer-wave infrared radiation is trapped by the atmosphere – predominantly the troposphere.[1] This heat retention, or 'radiative forcing', causes the natural greenhouse effect.

The troposphere of this planet is composed of approximately 78% nitrogen, 21% oxygen and 1% of the following: argon; traces of carbon dioxide; water vapour; methane; ammonia; hydrogen; and other minor gases. It is these latter gases – making up just the 1% – that are naturally occurring greenhouse gases, each absorbing a particular wavelength of infrared radiation, so trapping energy in the lower atmosphere and creating a heat blanket. Without this covering the earth's average surface temperature would be substantially lower than its present 15°C. Clearly, however, the greater the concentration of greenhouse gases, the thicker the blanket. So, while the natural effect keeps the earth around its life-promoting moderate temperature, manmade contributions to the greenhouse gases are likely to disturb this balance and heat up the atmosphere [2].

There are about 40 000 gigatonnes (Gt)[2] of naturally circulating carbon in the biosphere, 95% of which is in the deep oceans, and the rest in the atmosphere, plants, surface layers of oceans, topsoil, and a small fraction in animal life. An estimated massive 10 000 Gt of 'ancient' carbon is locked away in sedimentary limestone deposits and, crucially, in fossil fuels. The combustion by humans of these fossil fuels creates more carbon dioxide, water vapour and nitrous oxide (oxides of carbon, hydrogen and nitrogen), transforming the natural greenhouse effect into an unnatural one.

There are three other main anthropogenic greenhouse gases – methane and the entirely synthetic chlorofluorocarbons 11 and 12 (CFC-11 and CFC-12) – but carbon dioxide is the most significant because of its stability. Anthropogenic carbon dioxide emissions into the atmosphere total about 7 Gt each year, of which 5–6 Gt are from fossil fuel combustion and the remainder largely from the burning and clearance of forests. The biosphere's carbon cycle allows for removal of about 4 Gt a year through plant and ocean photosynthesis and solution, but this still leaves approximately 3 Gt of anthropogenic carbon dioxide left annually in the atmosphere [2].

Emissions of various greenhouse gases into the atmosphere have increased substantially since about 1800, and dramatically since 1950, essentially as the product of industrialisation. Seventy-five per cent of anthropogenic carbon dioxide comes from combustion of fossil fuels, especially coal, an increasing amount from motor vehicles, and the remainder largely from rainforest burning. Methane is mainly derived from irrigated agriculture, cows, mines, gas pipelines and rubbish tips; nitrous oxide comes from fossil fuel combustion and fertilisers.

Historically, there has been approximately a 100-fold increase in global energy use since 1800 and the same multiple in annual rate of carbon dioxide production. As a result, atmospheric carbon dioxide concentrations have increased by a third, half of that since the 1950s. Between 1800 and 1988, developed countries have been responsible for about 83.8% of industrial carbon dioxide emissions, 67.8% of total carbon dioxide emissions, and 66.9% of total combined carbon dioxide and methane (of which the USA contributes 33.2, 29.7 and 29.2%, respectively). In comparison, developing countries have been responsible for 16.2, 32.2 and 33.1%, respectively [3].

However, though developed countries are responsible for over four-fifths of historic carbon dioxide (in other words the total carbon dioxide now in the atmosphere), developing countries *currently* contribute 32% of annual global carbon dioxide emissions, expected to increase to around 44% by 2010 [4]. This is, of course, to do with industrialisation in many rapidly developing countries, for instance China. With regard to present-day emissions, developing countries tend to focus on per capita emissions, whereas industrialised countries point out that the total is what counts. These issues will be returned to later and in the following chapter.

According to most experts the increase in greenhouse gases in the atmosphere has resulted in a rise in surface temperature of the earth. In 1988 the UN Environment Programme and the World Meteorological Organization established a multidisciplinary body of over 300 scientists to advise governments, called the Intergovernmental Panel on Climate Change (IPCC). A 1990 report by the IPCC suggested that average global temperatures had risen by 0.3–0.6°C over the past 100 years, and predicted an average global increase of around 1°C by 2025 and 2.5–3°C by 2100. The approximate 1°C rise every 35 years would be greater at high latitudes and greater in the northern hemisphere [5]. Eleven years on, the 2001 IPCC report concludes that global average surface temperatures have increased by 0.6°C ± 0.6°C over the twentieth century, and are projected to rise by 1.4.–5.8°C by 2100 [6].

The consequences to human health of a rise in temperature and associated climatic changes are diverse, and can be grouped into direct and indirect effects [7]. Most of these health effects are projections based on empirical studies and expert judgement. New research methods, such as (scenario-based) predictive modelling, can be used to quantify the attributable effects; such techniques bring together different academic disciplines and approaches, and are accompanied by elements of uncertainty. The direct health effects result from greater exposure to thermal extremes, such as changes in mortality and morbidity from heat waves. Increased heat waves, exacerbated by increased humidity and urban air pollution, impact greatest on the elderly, the sick and those without access to air-conditioning. Other

Figure 9.1   Flooding: a couple in a boat are setting fish-traps in a river in Jianqzu Province, China, in 1999. The area is regularly flooded due to heavy rainfall, allowing freshwater snails hosting the blood flukes that cause the disease schistosomiasis to re-invade previously cleared land. Adverse weather events such as floods are expected to increase as a result of global warming (A. Crump, TDR, WHO/Science Photo Library).

direct effects are mediated through extreme weather events such as floods and storms, with their associated deaths, injuries, psychological disorders and infectious diseases [2].

Indirect effects stem from disturbances to complex ecological systems. Changes in the ranges and activity of vectors and infective parasites – through altered rainfall and temperature – are predicted to affect the geographical range and expansion of associated diseases such as malaria, dengue fever, trypanosomiasis and viral encephalitides. Altered local ecology will impact on water-borne and food-borne infective agents, causing increased incidence of gastrointestinal and other infectious diseases (Fig. 9.1). The level of impact will depend on the baseline incidence of such diseases, and will be compounded by factors such as floods and damage to public health infrastructures [8, 9].

Though hard to quantify, changed food productivity – especially crops – may result in malnutrition, hunger, impaired child development and growth, with increased morbidity and mortality. Tropical and subtropical countries will be the worst affected as the 'poor and economically underdeveloped populations...would be unable to offset agricultural yields by trade' [10]. Decreased water availability is expected for many populations in water-scarce regions, especially the subtropics. Just under a third of the world's population, 1.7 billion people, currently live in countries considered to be 'water-stressed' and this is expected to increase to 5 billion by 2025 [6].

The IPCC anticipates a globally averaged sea-level rise of 0.09 to 0.88 m by the end of this century with an associated rise in population displacement, damage to infrastructure, psychological morbidity, and problems with disposal of sewage and waste. Half the world's population lives within 60 km of the sea, and rising waters would particularly affect those living near coasts, on small islands, and those with limited material resources. (The Indian Ocean tsunami of 26 December 2004, although not a man-made event, demonstrated the devastation that water damage can cause to coastal areas.) Many fish populations will be put at risk from the sea-level and temperature rises rendering habitat unsuitable, and land-use changes are creating obstacles to migration [6].

The indirect impacts of climate change also include the biological effects of air pollution changes (including alterations in pollen and spore formation), such as morbidity and mortality associated with acute and chronic respiratory disorders and allergic conditions. Lastly, there are the varied public health consequences of social, economic and demographic dislocation.

As has been mentioned, the poorest countries are likely to be most heavily affected by the health effects of climate change, as they lack the resources to adapt accordingly. The story, however, is somewhat different in the UK. Warmer winters will be associated with less cold-related deaths. On the other hand warmer, and probably drier, summers will increase heat-related deaths and hospital admissions, but by a much smaller dimension; extra cases of food poisoning may occur. It has been predicted that more outdoor activity in the warmer weather will result in additional cases of skin cancer and cataracts. Within decades indigenous malaria may have become re-established in the UK, but probably only associated with the less threatening *Plasmodium vivax*. Sea-level rises and increased frequency of winter storms and gales will make flooding of low-lying coastal areas more likely. Most air pollutant levels are expected to decrease [11, 12].

Climate change and public health philosophy

From the perspective of public health philosophy, what is fascinating about climate change is that it throws open three new aetiological dimensions to population health and disease. First, the causes of greenhouse warming, and the resultant climate change and its health effects are anthropogenic. Excluding 'lifestyle' diseases – which an individual predisposes him- or herself to through personal activity[3] – there are plenty of examples of illnesses created by human activity, such as occupational cancers or the passive smoking example mentioned earlier. But in these situations the populations that create the environmental hazard predominantly experience the consequences.

What is different about climate change is that certain communities (and the individuals within them), through their adopted activities, will affect the health of other communities that may well not have taken up such activities. This opens up interesting and hitherto unexplored questions about personal responsibility, and also about the relationship between public responsibility and how this is expressed through policies such as those concerning public health. In other words, how do such responsibilities fit into the public health philosophy and practice of the perpetrating communities?

Second, the health effects of climate change are, to a substantial degree, likely to impact at a large geographical distance from their source. Aside from the equity issues relating to the *differential* impact and ability to mitigate or adapt accordingly – which are looked at in the next chapter – it is difficult to think of any other example[4] in which the activities of one community could so connectedly affect the health of a distant population. War is perhaps the closest parallel. Related to this point, the third new aetiological dimension that climate change throws up is that the health impacts of current (and past) activities will likely be the burden of generations to come. Once again, it is difficult to recall any similar example in the history of public health. So the question arises again of how do these spatial and chronological dimensions fit into the public health philosophy and practice of the perpetrating communities?

In some ways this takes us back to Part III in which it was argued that Western epidemiology and public health lack a coherent theoretical philosophy. But a different tack will be taken here, which looks more directly at the ethical foundations of public health. As with much public policy, public health is informed heavily by one moral theory, utilitarianism. Yet utilitarianism is problematic and seems out of touch with some of the world's current problems. As a guide for both personal and public morality, traditional utilitarianism appears in some important ways inappropriate. Indeed, the roots of the fresh dimensions to the health effects of climate change can be traced to the deficiencies of utilitarian theory.

So the second half of this chapter looks specifically at utilitarianism, its moral limitations and the relevance of these to climate change and public health philosophy. After that, Chapter 10 looks at the major challenging moral framework for public health (based on John Rawls's vision of social justice), as well as some alternative ideas. But first to utilitarianism.

Utilitarianism, climate change and public health

Utilitarianism falls into the consequentialist class of moral theories, in which the rightness or wrongness of an action or rule is determined by the consequences of that action or rule. The other main class of moral theories comprises deontological theories,[5] in which inherent characteristics of actions are of moral relevance, rather than an action's consequences. These are duty-based theories and include Kantianism, religious philosophies and the ethics of natural law.

Despite relentless ongoing criticism utilitarianism has proved a remarkably tenacious moral theory, the corner-stone to liberal democracy; both its persistence in, and significance to, Western political philosophy inevitably tie utilitarianism to ethical issues in public health. Utilitarianism became applied politically in the eighteenth century, and is most famously associated with Jeremy Bentham (1748–1832) and John Stuart Mill (1806–73). But utilitarianism had antecedents, and the main tenets of the theory were laid down earlier by philosophers such as John Locke (1632–1704) and David Hume (1711–76). Much of what Bentham and Mill had to say was not particularly new.

A shy individual, Bentham was – a little ironically – prone to depression. He preferred to remain outside the public eye, and became a founding member of the Philosophical Radicals only late in life. As Bertrand Russell pointed out, this historically enduring group did not have an especially strong philosophical bent and was not particularly radical. Their members did, however, unintentionally pave the way for the truly radical Socialists, from whom Bentham was keen to keep a safe distance, both theoretically and politically [13].

Bentham was a lawyer and was most interested in the relevance of his ideas to legislature. This element connected him strongly to one of his followers, Edwin Chadwick, because of a shared belief in improving the lot of those worst off through reform. But Bentham's concept of equality was strikingly at odds with that of certain successors, such as Marx and Engels, who provided a very different explanation for the historical processes that determine how inequalities arise, and what should be done to redress them. To Bentham, equality formed the basis of a calculus[6] in which each individual counted the same, and was the anchor of his utilitarianism.

Bentham's theory was founded on two linked principles: the principle of association and the principle of utility. The principle of association was a deterministic account of linked mental occurrences, akin to the modern 'conditioned reflex' but without the physiology. The principle of utility, or the greatest-happiness principle, is, however, what Bentham is best known for and rests on the premise that what is good is pleasure and what is bad is pain. Bentham came to this position through the belief, set out in the very first paragraph of *An Introduction to the Principles of Morals and Legislation* of 1789 [14], that human beings are subject to, and slaves to, two poles of sensation:

Nature has placed mankind under the governance of two sovereign masters, pain and pleasure . . . They govern us in all we do, in all we say, in all we think: every effort we can make to throw off our subjection will serve but to demonstrate and confirm it. In words a man may pretend to abjure their empire, but in reality he will remain subject to it all the while.

Extraordinarily, Bentham came up with 58 synonyms for pleasure, all denoting the same sensation, and famously remarked that 'Quantity of pleasure being equal, pushpin is as good as poetry.' Bentham claimed both psychological hedonism[7] and ethical hedonism[8] to be true. His principle of utility 'approves or disapproves of every action whatsoever, according to the tendency which it appears to . . . augment or diminish the happiness of the party whose interest is in question; or what is the same thing in other words, to promote or to oppose that happiness' [15].

Extrapolated from the individual to the larger, social domain, the principle of utility states that 'the greatest happiness of all those whose interest is in question . . . [is] . . . the only right and proper and universally desirable end of human conduct' [14]. So, one set of affairs is better than another if there is a greater balance of pleasure over pain, or a smaller balance of pain over pleasure. Empiricism was thus brought firmly into the foreground, as the right action could – in theory at least – be determined by summing up individual experiences of these two sensations. This process of quantification was Bentham's felicific calculus,[9] in which the 'audience' to be considered comprised all those affected by the action, each counting equally. Animals were not excluded from the calculus, as Bentham believed 'The question is not, Can they *reason*? Nor Can they *talk* but, *Can they suffer*?'

Alongside Bentham in the Philosophical Radicals was his friend James Mill, the Scottish philosopher, historian and economist, to whom Bentham lent money and gave a house. His son John Stuart Mill followed, to some extent, in his father's footsteps as a social reformer and advocate of utilitarianism. In his 1863 book *Utilitarianism* John Stuart Mill, like Bentham, extended moral consideration to the whole of sentient creation but, differently, made qualitative distinctions between pleasures. Mill felt that 'pleasures of the intellect, of the feeling and imagination, and of the moral sentiments' had higher value than those of mere sensation, so contrasting his non-hedonic, or ideal, utilitarianism with Bentham's hedonic version [16]. To Mill, it was better to be a dissatisfied Socrates than a satisfied fool. But, in general respects, Mill was in accord with Bentham, and both equated pleasure with happiness and pain with unhappiness [17]:[10]

The creed which accepts as the foundation of morals, Utility or the Greatest Happiness Principle, holds that actions are right in proportion as they tend to promote happiness, wrong as they tend to produce the worst of happiness. By happiness is intended pleasure and the absence of pain; by unhappiness, pain, and the privation of pleasure.

Although critiques of modern utilitarianism, and their relevance to public health, will be looked at a little later, it is necessary to point out here a serious problem with the theory's early forms. That is, utilitarianism as depicted by both Bentham and Mill, makes an erroneous conceptual leap of inferring from what 'is' in the world to what 'ought' to be. This move from description of fact to moral prescription has been termed 'the naturalistic fallacy' by Moore [18].

The naturalistic fallacy can be taken in different ways with regard to utilitarianism. One possibility is that ethics can be debunked, and science allowed to guide human action. Roger Scruton articulates (but does not agree with) this in commenting that provided pleasures and pains 'are thought of as bearing only a quantitative, but not a qualitative, relation to each other (that is, provided the principle aim and subject matter of ethics is forgotten), then it is possible to envisage a solution to all moral problems' [19]. Bertrand Russell [20] is more straightforwardly damning:

> John Stuart Mill, in his Utilitarianism, offers an argument which is so fallacious that it is hard to understand how he can have thought it valid. He says: Pleasure is the only thing desired; therefore pleasure is the only thing desirable. He argues that the only things visible are things seen . . . and similarly the only things desirable are things desired. He does not notice that a thing is 'visible' if it can be seen, but 'desirable' if it ought to be desired. Thus desirable is a word presupposing an ethical theory; we cannot infer what is desirable from what is desired.

If indeed this fallacy *is* a fallacy, it is perhaps surprising that utilitarianism has endured. Yet few would demur that utilitarianism strongly underpins much of contemporary moral and political thinking and action. Good national policies are judged to be those that increase overall wealth, the modern euphemism for the greatest happiness, and good public health policies are judged to be those that demonstrably improve population health. These two examples, however, also capture one way in which utilitarian beliefs have changed with time – the separation of the private and the public.

For the early followers, there was no division. Utilitarianism, as a moral theory, provided a guide to both personal behaviour and also to public decision-making. But today it is quite acceptable – and perhaps even the norm – for individuals to draw on various ethical ideas to inform their private behaviour, while expecting governments, public bodies and institutions essentially to act for the common good. While those private ethical ideas tend to have their roots in duty-based theories or, for some, religious beliefs, the notion of the common good is undeniably utilitarian. This is the moral pluralism of contemporary Western, largely secular, societies.

That is not to say that individuals in such societies do not seek to improve their own happiness, in fact quite the opposite. But the pursuit of happiness is removed from moral consideration, and has become something closer to being taken for granted, a lifelong endeavour shaped by society, unquestionably accepted and followed. The inevitably elusive chase finds happiness disguised as, *inter alia*, 'healthism', obsession with risk aversion, and consumerism [21, 22]. Not surprisingly, therefore, faced with the lifestyle changes that might be required to offset climate change and its global health effects, most individuals are not keen on sacrificing, to any great extent, pursuit of their own happiness-oriented goals, despite token environmental soundings to the contrary. Affordable mitigation strategies would be preferable.

Picking up on this the philosopher Alisdair MacIntyre has put the blame for today's moral ambivalence squarely on the shoulders of utilitarianism and the selfishness it has engendered. He argues that 'the individualism of modern society and the increasingly rapid and disruptive rate of social change brings about a situation in which for increasing numbers there is no over-all shape to the moral life but only a set of apparently arbitrary principles inherited from a variety of sources.' In such circumstances, he continues, 'the need for a public criterion for use in settling moral and evaluative disagreements and conflicts becomes even more urgent and ever more difficult to meet'. He suggests that the utilitarian criterion, which appears to embody the liberal ideal of happiness, is apparently without rivals, 'and the fact that the concept of happiness which it embodies is so amorphous and so adaptable makes it not less but more welcome to those who look for a court of appeal on evaluative questions which they can be assured will decide in their own favour' [23].

Further, MacIntyre holds the early utilitarians directly responsible for today's woes [24], and emphatically questions the price to be paid:

But it is necessary to emphasize that the utilitarian advocacy of the criterion of public happiness is not only a mistake. That it seems so obviously the criterion to be considered in certain areas of life is something we owe to Bentham and Mill.

The concept of happiness is, however, morally dangerous in another way; for we are by now well aware of the malleability of human beings, of the fact that they can be conditioned in a variety of ways into the acceptance of, and satisfaction with, almost anything. That men are happy with their lot never entails that their lot is what it ought to be. For the question can always be raised of how great the price is that is being paid for the happiness. So the concept of the greatest happiness of the greatest number could be used to defend any paternalistic or totalitarian society in which the price paid for happiness is the freedom of the individuals in that society to make their own choices.

This now brings us back to the ways in which utilitarianism has changed from its early forms. In the private sphere happiness has become morally detached, confusingly and ambiguously entangled with the ethos of Western self-centredness. But in the public sphere utilitarian theory has developed. As theorists began to recognise that summation of happiness was, not only practically difficult, but also an insufficient and incomplete reflection of human goals and needs, alternatives were sought.

Some utilitarian theories today have moved from pleasure and pain as criteria of the consequences of acts, rules or policies to rather different criteria: preferences (preference utilitarianism) or interests (welfare utilitarianism). The idea of human preferences has been felt to reflect more realistically that individuals have different desires and wants, and it is the satisfaction of those preferences that is important, and ultimately is a source of utility for any person. Quantitative analysis of people's preferences is, of course, possible and is partly reflected in the market-place, making preference satisfaction appealing to economists. One state of affairs is better than another if more people's preferences are satisfied or if there are less people with unsatisfied preferences. The dramatic culture shift in the NHS towards providing patients with choice is a contemporary example of this [25].

But this of course leads again to the question of what determines preferences, and so back to MacIntyre's criticisms. Another modern expression of utilitarianism deals, instead, with human interests, with interests referring to basic needs, the satisfaction of which allows people to move on to attaining their preferences. Health, shelter and basic sustenance are examples of interests, and welfare economics is the academic and political enterprise in which utilitarian attempts to maximise interests are analysed and expressed. Significantly, welfare utilitarianism has given the intuitively appealing notion of utility some practical content [26].

It is not appropriate to look further here at welfare economics, but sufficient for the scope of this chapter to recognise the recent developments in utilitarianism, and also to mention the deficient assumptions in summing and aggregating utilities (as interests or preferences) into an overall measure – assumptions about the comparability of goods and people (the utility to me of a one-unit increase in wealth or health may be very different to the utility to another person of the same increase).

With a broad picture of utilitarianism in mind, it is now appropriate to look at criticisms of utilitarian philosophy and how these relate to climate change and public health.

Critiques of utilitarianism and relevance to climate change

The first criticism of utilitarianism is that the theory, in its classical or present economic form, necessitates the enumeration and summation of utilities in some shape or form. Utilitarianism then uses the results of this process as the moral basis to guide actions or policies. In its classical form an obvious difficulty was how to quantify happiness, along with the problem outlined earlier of whether happiness is an appropriate moral goal in the first place. Preference- and welfare-based utilitarianism circumvent the latter issue but do not get around the issue of quantification.

In fact, modern versions of utilitarianism do precisely the opposite. They are reliant, perhaps more than ever, on empirically obtained quantitative information as the basis for acting. They place, metaphorically, all the moral eggs in the basket of a positivist conception of science. In a classic contemporary book containing essays for and against utilitarianism, the philosopher Bernard Williams describes contemptuously the appeal of utilitarianism in that it picks up 'little of the world's moral luggage', preferring instead to place huge demands on information because 'even insuperable technical difficulty is preferable to moral unclarity, no doubt because it is less alarming' [27].

This appeal may be economically and politically advantageous, in the short-term at least, but it (or perhaps because it) raises almost insuperable problems for climate change. There may be a general consensus now on the scientific proof that climate change is actually happening, as outlined earlier, but there is no agreement about what should be done about it. This can be conveniently deferred to disagreeing scientists, working together across disciplines in ways not previously demanded, using novel techniques with inherent uncertainties.

The same problem exists for assessing the health impacts of climate change. They are based on, at best, plausible predictions using models and methods (including expert judgement and inference) developed because of the newness of the topic and the lack of alternatives. But utilitarian calculations prefer concrete facts to ranges and possibilities. This provides an easy escape route for policy-makers but also reflects the compounded difficulties of comparing or trading utilities in the climate change debate. With both health and monetary impacts fraught with empirical impracticabilities, weighing up options and alternatives is extremely hard.

More subtly, it raises the important point that some elements are more amenable to scientific enquiry and analysis than others. The environment, for example, may be less amenable in some ways. For instance, it is not easy to place a utility function on the value individuals may, or may not, place on retaining a beautiful area of wilderness or an unpolluted atmosphere. This will be returned to later but it is a clearer task to calculate the economic costs of climate change and the mitigation strategies to prevent it, than to reliably quantify the health impacts or value of the environment, so creating a bias in areas of consideration; and this does not even touch the question of how to compare different utilities. This is a fundamental issue recognised by Williams [28]:

For to exercise utilitarian methods on things which at least seem to respond to them is not merely to provide a benefit in some areas which one cannot provide in all. It is, at least very often, to provide those things with prestige, to give them an unjustifiably large role in the decision, and to dismiss to a greater distance those things which do not respond to the same methods. Just as in the natural sciences, scientific questions get asked in those areas where experimental techniques exist for answering them, so in the very different matter of political and social decision weight will be put on those considerations which respected intellectual techniques can seem, or at least promise, to handle.

The second criticism of utilitarianism, and its framing of climate change policies, relates to proximity. As has been described, classical and modern versions of utilitarianism involve quantification and summation of individual utilities, whether happiness, preferences or interests. But who should be included in the arithmetic? Bentham and Mill predicated that the pleasures and pain of all affected by the action, the audience, should be considered, including – to a lesser degree – nonhuman animals. Leaving animals aside for now, determining which humans might be affected is no easy task.

Although circumstances in the nineteenth and twentieth centuries were more contained than today by the technology of the time, the same concentration on individuals (the audience members) close in space and time applies to both eras. This is because, once again, the utilitarian calculus favours consideration of that over which there is greater certainty. The philosopher Robert Goodin highlights that utilitarians may want to include the utilities of all those affected by an action in any given calculation [29], but in practice it is unlikely:

. . . utilitarians can go on to say, perfectly properly, that as a purely pragmatic matter their calculations will often lead us to show some apparent favouritism toward those near and dear to us. It is easier to know what people nearby need, and how best we can help; . . . Those are purely contingent, pragmatic considerations, to be sure. In the ideal world, they may be absent. But in the real world, they are powerfully present.

This creates special problems for policies relating to climate change. At the national level, and at the local level within countries, policies usually take into account their effects on individuals contained by their boundaries. Climate change would appear to open up the borders by demanding that those from afar are considered too. But it is hard at present to know how to incorporate such requirements, and it remains difficult to believe that such tough decisions will be made by politicians with national, party and their own interests at heart. The limited concessions to date in the high-profile international climate change meetings affirm the somewhat bleak outlook, as is discussed further in Chapter 10.

Similarly, public health policies within, say, a health district in England or Wales, would need to take heed of their distant impacts. As recent authors in the global bioethics literature have put forward, Western hospitals are significant producers of greenhouse gas emissions and should be making every effort to substantially reduce these, so as to avoid the irony of a healthcare system in one country adding to the healthcare problems of another [30]. Yet it is difficult to envisage the dramatic changes needed, and token efforts are likely instead. This is because at the heart of that very problem lies the unrealistic notion of health that is a central feature of Western living, and the corresponding reliance that has developed on healthcare services. A different example, that of speed bumps, though relatively unimportant in terms of climate policy, illustrates the proximate utilitarian spirit within public health: speed bumps are largely popular because they reduce traffic accidents locally but they may also increase greenhouse gas emissions from motor vehicles.

So far the focus of this second criticism has related to geographical proximity but utilitarianism also has a temporal bias. The utilitarian philosopher J. J. C. Smart argues that it is impossible to envisage the total future situation because it stretches to infinity [31]. According to Smart it is unnecessary in practice to consider very distant consequences as these in the end approximate rapidly to zero like the furthermost ripples on a pond after a stone has been dropped into it. He defends this presentism [32]:

The necessity for the 'ripples in the pond' postulate comes from the fact that usually we do not know whether remote consequences will be good or bad. Therefore we cannot know what to do unless we can assume that remote consequences can be left out of the account.

This issue is particularly acute for climate change, and policies related to it, as the environmental, financial and health impacts will not only occur in the future but also in the distant future. Economists have a general way of dealing with this phenomenon called 'discounting', an analytical tool to compare economic effects that occur at different points in time.[11] But there are different discount rates available and 'the choice of discount rate is of crucial technical importance for analyses of climate change policy, because the time horizon is extremely long, and mitigation costs tend to come much earlier than the benefits of avoiding damages' [33].

There has been extensive unresolved debate about discounting in assessment of climate change policies, some authors challenging the traditional view and showing that early attention to abatement will minimise risk to both environmental and economic systems [34]. The debate, however, reminds us that facts alone cannot provide moral judgements. The 2001 IPCC publication emphasises that uncertainty regarding the discount rate 'relates not to calculation of its effects, which is mathematically precise, but to a value judgement about the appropriateness of the present generation valuing services for future generations' [35]. Environmental philosophers have pointed out that any form of discounting devalues the environment and the benefits it holds for future generations.

The final criticism of utilitarianism relates to equity. The summation and averaging of utilitarian calculations insufficiently recognises the importance of how utilities are distributed within the population under consideration. Whether the utility is health or wealth, there may be no difference in the aggregate amount of the utility between a population in which a small number have a lot of (good) health and the remainder have poor health, and a population in which everyone is reasonably healthy. This does not sit comfortably with our common-sense morality, as Williams states [36]:

> In this light, utilitarianism does emerge as absurdly primitive, and it is much too late in the day to be told that questions of equitable or inequitable distribution do not matter because utilitarianism has no satisfactory way of making them matter. On the criterion of maximising average utility, there is nothing to choose between any two states of society which involve the same number of people sharing in the same aggregate amount of utility, even if one of them is relatively evenly distributed, while in the other a very small number have a very good deal of it; and it is just silly to say that in fact there is nothing to choose here.

So, if climate change illustrates that utilitarianism has some drawbacks as an ethical framework for public health policies, it may be that a different approach is needed. Here, recent developments in the climate change debate suggest an alternative might be emerging.

NOTES

1. The earth's atmosphere extends approximately 100 km from the earth's surface. It is comprised of the 10 km troposphere closest to the earth, which contains 90% of the mass of the atmosphere, then the stratosphere band between 10 and 50 km, which is less dense and contains the ozone 'layer', and finally the mesosphere.
2. One gigatonne (Gt) = one billion (1 000 000 000) tonnes.
3. Personal choice, however, such as the ability to stop smoking, may be affected by factors such as employment status and social support, both of which are linked to deprivation.
4. Apart from economic activity, which is related to human lifestyles.
5. Challenges to this supposition are presented in Chapter 10.
6. The calculus referred to a calculation rather than the modern understanding as a particular method in mathematics.
7. Psychological hedonism argues that human beings are (psychologically) driven to strive for pleasure.
8. Ethical hedonism argues that it is morally right that human beings should strive for pleasure.
9. Also known as the optimific or hedonic calculus.
10. This is a 'stipulative definition' and can be resisted.
11. The basic premise behind discounting is that a million pounds to me now is of more value than a million pounds in a year's time.

REFERENCES

1. J. M. Last, Global change: ozone depletion, greenhouse warming, and public health. *Ann. Rev. Pub. Health*, **14** (1993), 123.

2. A. J. McMichael, *Planetary Overload: Global environmental change and the health of the human species* (Cambridge: Cambridge University Press, 1995), pp. 132–73.

3. IPCC, *Climate Change 1995: Economic and social dimensions of climate change* (Contributions of Working Group III to the Second Intergovernmental Panel on Climate Change (Cambridge: Cambridge University Press, 1996).

4. C. Hamilton, Justice, the market and climate change. In N. Low, ed., *Global Ethics and Environment* (London: Routledge, 1999), pp. 90–105.

5. J. T. Houghton, G. J. Jenkins and J. J. Ephraums, *Climate Change. The IPCC assessment* (Cambridge: Cambridge University Press, 1990).

6. IPCC, *Climate Change 2001: Impacts, adaptation, and vulnerability* (Cambridge: Cambridge University Press, for IPCC, 2001).

7. A. J. McMichael, *Global Environmental Change and Human Health*. Paper presented at seminar on Global Changes and Human Health, Royal Swedish Academy of Sciences, Stockholm, 29 May 1996.

8. A. J. McMichael and A. Haines, Global climate change: the potential effects on health. *BMJ*, **315** (1997), 805–9.

9. P. Epstein, Emerging diseases and ecosystem health: new threats to public health. *Am. J. Pub. Health*, **85** (1995), 168–72.

10. A. J. McMichael and A. Haines, Global climate change: the potential effects on health. *BMJ*, **315** (1997), 808.

11. Department of Health, *Health effects of climate change in the UK* (consultation document) (London: HMSO, 2001).

12. A. J. McMichael and A. Haines, Global climate change: implications for research, monitoring, and policy. *BMJ*, **315** (1997), 870–4.

13. B. Russell, *History of Western Philosophy* (London: Routledge, 1991). [Originally published in 1946.]

14. Quoted in R. Scruton, *A Short History of Modern Philosophy* (London: Routledge, 1996), p. 224.

15. Quoted in J. Rachels, *The Elements of Moral Philosophy* (New York: McGraw-Hill, 1993), p. 91.

16. C. E. Harris, Jr., *Applying Moral Theories* (Belmont: Wadsworth, 1997), p. 128.

17. Quoted in A. Flew, ed., *A Dictionary of Philosophy* (London: Pan Books, 1984), p. 361.

18. G. E. Moore, *Principia Ethica* (Cambridge: Cambridge University Press, 1903).

19. R. Scruton, *A Short History of Modern Philosophy* (London: Routledge, 1996), pp. 224–5.

20. B. Russell, *History of Western Philosophy* (London: Routledge, 1991), p. 744. [Originally published in 1946.]

21. O. H. Forde, Is imposing risk awareness cultural imperialism? *Soc. Sci. Med.*, **47**:9 (1998), 1155–9.

22. D. Porter, *Health, Civilisation and the State* (London: Routledge, 1999).

23. A. MacIntyre, *A Short History of Ethics* (London: Routledge, 1989), p. 243.
24. *Ibid.*, pp. 237–8.
25. Department of Health, *The NHS Plan* (London: HMSO, 2003).
26. R. E. Goodin, Utility and the good. In P. Singer, ed., *A Companion to Ethics* (Oxford: Blackwell, 1997), pp. 241–8.
27. B. Williams, A critique of utilitarianism. In J. J. C. Smart and B. Williams, *Utilitarianism: For and against* (Cambridge: Cambridge University Press, 1991), p. 137. [Originally published in 1963.]
28. *Ibid.*, p. 148.
29. R. E. Goodin, Utility and the good. In P. Singer, ed., *A Companion to Ethics* (Oxford: Blackwell, 1997), p. 247.
30. A. Jameton, Outline of the ethical implications of the earth's limits for health care. *J. Med. Hum.*, **23**: 1 (2002), 43–59.
31. J. J. C. Smart, An outline of a system of utilitarian ethics. In J. J. C. Smart and B. Williams, *Utilitarianism: For and against* (Cambridge: Cambridge University Press, 1991), pp. 3–74. [Originally published in 1963.]
32. *Ibid.*, pp. 33–4.
33. IPCC, *Climate Change 1995: Economic and social dimensions of climate change* (Cambridge: Cambridge University Press, for IPCC, 1996), p. 8.
34. M. Ha-Duong, M. J. Grubb and J.-C. Hourcade, Influence of socioeconomic inertia and uncertainty on optimal CO_2-emission abatement. *Nature*, **390** (1997), 270–3.
35. IPCC, *Climate Change 2001: Impacts, adaptation, and vulnerability* (Cambridge: Cambridge University Press, for IPCC, 2001), p. 97.
36. B. Williams, A critique of utilitarianism. In J. J. C. Smart and B. Williams, *Utilitarianism: For and against* (Cambridge: Cambridge University Press, 1991), pp. 142–3. [Originally published in 1963.]

10

Climate change, social justice and other moral frameworks

In Part IV of this book we are looking at the latest conception of air and health, the approach to dealing with climate change. In Chapter 9 the basic science of greenhouse warming and climate change was first presented, and then utilitarianism was examined, both in relation to climate change, and as the ethical theory underpinning public health more generally. Criticisms of utilitarianism were discussed. This chapter explores social justice as a challenge to utilitarianism, and the relevance of social justice to climate change and public health. Other moral frameworks, or approaches, are also considered.

Social justice, climate change and public health

There is a huge literature on justice, stretching back as far as the Greeks. Aristotle, for instance, in the *Nicomachean Ethics* considers just actions, and likens the characteristic of being just to the other 'excellences' – or virtues of character. For Aristotle, justice is a mean, injustice represents the extremes, and the just man[1] recognises how to determine an individual's appropriate share. Justice is 'that by which the just man is said to do by choice what is just and to be one who will distribute either between himself and another or between two others . . . so as to give what is proportionately fair' [1]. Aristotle's perception of justice, and some of its problems, are returned to later in this chapter.

In recent times social justice has come to embody aspects of the last part of Aristotle's definition: fairness and proportionality. In contrast with legal and retributive justice, social justice is about the distribution of society's benefits and burdens, and the sociopolitical mechanisms that enable such distribution to occur. This distinction of degree is exemplified in the debate about health care provision [2, 3]. Despite extensive ethical and philosophical discussion about fair allocation (or rationing) of healthcare resources and services, in practice economic analyses have dominated decision-making, and it is questionable how much the debate has actually influenced the institutions and processes that determine decisions at national and local levels. Only recently has research begun to look more closely at these issues [4].

The latest conception of air and health – the approach taken in dealing with climate change – has certainly extended the boundaries of moral debate in areas of public policy and public health policy. Because the causes and effects of climate change are differentially distributed, the reasonableness of basing decisions purely on utilitarian economic thinking has been questioned. Climate change may have pushed to the forefront, and into the public realm, issues that appeal directly, and instinctively to people's common-sense morality. It simply does not seem fair that islanders in the South Pacific should lose their homes because of two centuries of industrialisation in the West, and the profligate lifestyles this development has engendered.

As a result there has been a flurry of academic work looking at equity considerations in the climate change debate. Before outlining what this work has explored, it is necessary to consider what frameworks could be used. One way of determining how to distribute the costs of climate change mitigation and adaptation policies would be *not* to try to 'falsely' distribute such costs at all but to allow the market to decide. But libertarian, or market utilitarian, approaches would most likely lead to rich countries not valuing, or not wanting to pay, for such policies, and poor countries being unable to afford them. Unless there was some kind of catastrophic threat from climate change, poor countries might well be left to simply deal with the consequences.

An alternative framework would be contractarian, also sometimes called administrative utilitarian. In this approach, the limits of using total sum or average utility as a sufficient determinant of policy are acknowledged, and efforts are made to incorporate additional dimensions to economic calculations to allow for more informed, and hopefully fairer, distribution. These have figured strongly in the debate and will be returned to, but criticisms can still be levelled at inherent deficiencies in valuing human health, the environment, communities at a distance, and future generations.

The third framework has been strongly advocated as egalitarian, and has drawn on the ideas of social justice. In fact, one person's theory has been stressed within this dimension of the debate, John Rawls, whose name is scattered among the articles, discussion papers, and policy-related documents on equity issues in climate change. But Rawls's work is also emphasised by those working on the contractarian approach mentioned above, so, before examining how Rawls's ideas have been applied, it is necessary to look at the theory itself.

John Rawls's theory of justice

Rawls's *A Theory of Justice* [5] was first published in 1971 and 30 years on the impact remains remarkable. Despite criticisms, revisions and reprints, arguably no rival theory of justice has contested its pole position nor lasted so well. Yet many public health workers today have probably never heard of it, a reflection that is considered in the recommendations section of the book.

Rawls thinks of justice as fairness. He starts from the premise that utilitarianism is an inadequate, inappropriate and ultimately unjust moral or politico-economic tool for making distributive decisions in society. For Rawls justice *denies* that 'the loss of freedom for some is made right by a greater good shared by others' and justice does *not* allow 'that the sacrifices imposed on a few are outweighed by the larger sum of advantages enjoyed by many' [6]. Instead Rawls defines justice as 'a characteristic set of principles for assigning basic rights and duties and for determining what they take to be the proper distribution of the benefits and burdens of social cooperation' [7].

Rawls makes two other key assertions. First, he argues that people's perceptions of entitlement – and so too of justice or fairness – are inevitably shaped by their own backgrounds, interests and social organisations. While Rawls accepts that human beings naturally have certain interests – for instance, striving for basic primary goals – most interests are not of this nature and any agreed notion of justice needs to be reached before the undue influence of unnatural interests. Second, he predicates that any social advantages obtained through chance – by birthright or natural endowment – are essentially unfair.

Putting these together Rawls sets out to establish the principles of justice for the basic structure of society that would be agreed by individuals in an 'original' (or abstract pre-existence) state. Taking the form of a social contract these principles are those that 'free and rational persons concerned to further their own interests would accept in an initial position of equality as defining the fundamental terms of their association' [8]. This initial, or original, position corresponds to the state of nature in the traditional theory of the social contract. Rawls purports that the 'original position is . . . the appropriate initial status quo, and the fundamental agreements reached in it are fair.' This, he continues, 'explains the propriety of the name 'justice as fairness'; it conveys the idea that the principles of justice are agreed to in an initial situation that is fair' [9].

So Rawls sets up this original position and makes procedural justice the basis of his theory: the procedure is the contract, or the principles of justice, that would be determined by those sitting in the hypothetical position. Rawls now recognises the importance of nullifying 'the effects of special contingencies which put men at odds and tempt them to exploit social and natural circumstances to their own advantage.' This he does by situating parties making the contract, and in the original position, behind 'a veil of ignorance' so that they 'do not know how the various alternatives will affect their own particular case and they are obliged to evaluate principles solely on the basis of general considerations' [10].

Behind this veil 'no one knows his place in society, his class position or social status; nor does he know his fortune in the distribution of natural assets and abilities, his intelligence and strength, . . . his conception of the good, the particulars of his rational life, or even the special features of his psychology such as his aversion to risk or liability to optimism or pessimism' [10]. Of special significance to the climate-change debate, Rawls also highlights the moral relevance of the environment and of future generations:

The persons in the original position have no information as to which generation they belong. These broader restrictions on knowledge are appropriate in part because questions of social justice arise between generations as well as within them, for example, the question of the . . . conservation of natural resources and the environment of nature . . . They must choose principles the consequences of which they are prepared to live with whatever generation they turn out to belong to.

So parties in the original position have facts concealed from them by the veil of ignorance. They do not know where they will fall in society, what ordinal levels of wealth and income they will receive, what opportunities will befall them by virtue of their social positions. They do not know what their lot will be. From this position Rawls argues that parties would agree to two principles of justice and, by extension, these are the principles that society should strive to promote and maintain. The two principles are:

1. Each person is to have an equal right to the most extensive scheme of basic liberties compatible with a similar scheme of liberties for others.
2. Social and economic inequalities are to be arranged so that they are both
    a. to the greatest expected benefit of the least advantaged and
    b. attached to positions and offices open to all under conditions of fair equality and opportunity [11].

The first principle sets out generally that all social values – liberty and opportunities, income and wealth, and the social bases of self-respect – are to be distributed equally. These Rawls calls primary *social* goods, assets that every rational human being is presumed to want but the attainment of which is unduly, and unfairly, influenced by historical and social fortune. In contrast Rawls labels individual talents and abilities, such as intelligence and vigour, as primary *natural* goods, the differential distribution of which is an acceptable aspect of the human condition. He acknowledges that these goods or characteristics are influenced by social structure but claims they are less directly under its control. Interestingly Rawls also brackets health as a natural good, a perhaps contentious point given the recent growing understanding of the relationship between income and health. This is touched on later in the section on inequalities in health.

Rawls is not concerned with challenging the allocation of natural goods but justice requires fair distribution of social goods. So the second principle, named the difference principle, attempts to rebalance the arbitrary effects of the natural lottery, and gives weight to considerations he describes under the principle of redress [12]:

This is the principle that undeserved inequalities call for redress; and since inequalities of birth and natural endowment are undeserved, these inequalities are to be somehow compensated for. Thus the principle holds that in order to treat all persons equally, to provide genuine equality of opportunity, society must give more attention to those with fewer native assets and to those born into the less favourable social positions. The idea is to redress the bias of contingencies in the direction of equality.

Rawls describes the two principles as providing together the 'maximin' (*maximum minimorum*) solution to problems of social justice. Any putative policy and alternatives will have a range of outcomes which may differentially impact on those well off and those less well off. The maximin rule ranks alternatives by their worst possible outcome and advises adopting the option, the worst outcome of which is superior to the outcome of the others. The worst outcome refers to the impact of the policy on those worst off, and the goal is to improve their situation maximally [13].

Rawls elaborates that the best arrangement possible, a perfectly just scheme, is when the expectations (or lot) of the least advantaged are maximised, and no change in those better off can improve the situation of the worst off. But society is constantly changing, equilibrium is not fixed, so a just arrangement (but not the perfectly just arrangement) is that in which any increase in the expectations (or lot) of the more advantaged would increase that of the least advantaged. Any improvement in the welfare of the more fortunate must contribute to the welfare of the less fortunate. But the perfectly just state has not been reached, and only exists when the lot of the worst off has been maximised and any improvement to the better off would not improve the lot of the worst off.

Rawls adds a vital caveat, which is often under-emphasised. The situation described above could guide policies that widened existing inequalities, in say wealth, so long as the poorest had some gain, however tiny. A policy that led to the richest 10% earning an extra £1 million could be acceptable if the poorest, as a result, earned an additional penny. Of course it would have to be ranked and considered against competing policy options but nevertheless would be theoretically possible. So Rawls qualifies the conditions and stresses that a scheme is unjust when the lot or expectations of the advantaged are 'excessive', and a decrease in their situation would improve that of the least favoured [14].

Fairness issues in the climate change debate

Armed with the basics of Rawls's theory of justice, it is now possible to return to the climate change debate. The starting-point for the distributive concerns in climate change are three related questions: who is responsible for the problem; who will suffer (most) from the problem, and how; and who will bear the costs of abatement? The four fairness issues in climate-change policy which correspond to these questions have been expressed as follows [15]:

1. What is a fair allocation of the costs of preventing the global warming that is still avoidable?
2. What is a fair allocation of the costs of coping with the social consequences of the global warming that will not, in fact, be avoided?
3. What background allocation of wealth would allow international bargaining (about the first two points) to be a fair process?
4. What is a fair allocation of greenhouse gases over the long-term and during transition to the long-term allocation?

As background to these four areas, the Australian government suggested that a number of key points inform discussion about equity issues. Some of these were alluded to in Chapter 9 but they have been conveniently synthesised by Hamilton [16]. First, the problem is one of the global commons: emissions in one country contribute to climate change in all, and each country should theoretically benefit from efforts of others to reduce emissions [17]. Second, the high concentration of greenhouse gases currently in the atmosphere is due mainly to developed countries, a by-product of the accumulation of industrial wealth. Third, poor countries will suffer most the consequences of climate change, especially those in the tropics. Industrialising countries, however (in particular China, Brazil and India), will become dominant emitters of greenhouse gases in two or three decades. Next, the major climatic effects are not expected until the middle of this century and onwards, but action is needed now. Last, there are significant uncertainties about the type, extent, nature and location of climate change, and its human impacts. There is one further fairness matter but that relates to the policy-making process.

With regard to the issues of distributional equity above, Rayner *et al.* argue that there are broadly three ways of allocating burdens (costs): libertarian (market libertarian); contractarian (administrative utilitarian); and egalitarian (anthropocentric and nature-centric) [15]. The distinction between the last two categories is important here, as it will be noted that earlier in the chapter Rawls was bracketed with an egalitarian approach, rather than a contractarian approach.

A libertarian approach, as previously mentioned, is favoured by those who hold that when damage from climate change is of sufficient degree, the market-place will place appropriate value on rectification, and scientific methods will develop technological fixes that will (hopefully) remedy the problems. Some of these – such as cleaner engines and cleaner fuels – are already underway, and historically science has been very successful in creating solutions to complex problems. In economic terms delay may also be cheaper, although this has been challenged [18]. There is serious concern, however, that procrastination in introducing firm action will result in a state being reached in which resolution is impossible – if indeed that is not already the case. The widely flouted 'precautionary principle' dictates that when there is serious doubt about likely environmental impacts and consequences, decisions should be made that err on the side of safety [19]. Also, mitigation does not necessarily need to be expensive.

In trying to address the fairness issues in climate change, debate has actually focused on the second approach, administrative utilitarian (or contractarian), drawing in, to a degree, some Rawlsian ideas of social justice. The IPCC, for example, distinguishes two categories of equity as significant to climate change analyses: procedural equity and consequentialist equity. The former is largely about making policy, focusing on the criteria and methods for implementing fair procedures for design of, and participation in, the decision-making processes, as well as respect for legal rights. It is about inclusion, fairness and openness at all stages in the policy-making processes.

Consequentialist equity, in contrast, corresponds to items listed earlier, and is about the outcomes of climate change (and policies addressing climate change): justice and fairness in respect of the *impacts* of climate change, and justice and fairness in respect of *abatement* – in other words the distribution of burdens and allocation of benefits associated with reducing greenhouse gas emissions and managing climate change. Consequentialist equity has been further divided into *intra*generational equity (although actions by individuals in contributing to greenhouse gas emissions may affect anyone, impacts reflect vulnerability and are borne differentially by social groups or countries depending on their geography, economic development and so forth) and *inter*generational equity (costs of abatement may be borne now but benefits may not be realised well into the future).

The IPCC has postulated that there are several traditions in attempting these calculations, some of which have an egalitarian base. The traditions are, in the main, about how the cake is divided rather than how the cake is made, perceived or valued: 'parity' of burdens and benefits (equal distribution to all claimants, 'egalitarian'); 'proportionality' of burdens and benefits (distribution in proportion to contribution of claimants); priority (according to greatest need, emphasises basic needs and minimum level of well-being); classical utilitarianism (maximising total utility); and Rawlsian distributive justice (described as equal distribution unless unequal distribution operates to benefit the least advantaged). The IPCC [20] sometimes appears to consider Rawls's ideas as central to their approach:

A basic needs approach . . . involves allowing countries the right to emit the minimum levels of greenhouse gases needed to meet the *basic needs* of their citizens, defined as the minimum consumption levels needed to support full participation in society, and then requiring countries to buy (or pay taxes on) the rights to emission levels above these . . . This approach can be related to Rawlsian philosophy.

In general, however, the IPCC describes different traditions, drawing on a variety of relevant criteria, with Rawls invoked sporadically, as shown in Box 10.1.

Box 10.1 Possible ways of distributing the burdens and benefits of climate change as described by the International Panel on Climate Change

A. *Differences between countries or regions* based on:
   i. wealth/consumption
   ii. patterns of greenhouse gas emissions – historical (national debt concept), current, or future emissions
   iii. vulnerability to impacts
   iv. capacity to deal with changes
   v. resources.

B. *Distributing costs of coping* based on:
   i. paying for costs of coping based largely on existing wealth
   ii. sharing risks (Rawls's maximin, precautionary selection of the strategy minimising the worst outcome)
   iii. policy implications of the costs of risk-bearing.

C. *Distributing future emissions and abatement costs* based on:
   i. egalitarianism
   ii. Rawlsian criteria and basic needs, allowing the right to emit to meet basic needs, then buy or pay taxes above threshold
   iii. proportionality and 'polluter pays' principle
   iv. historical contribution
   v. comparable burdens and ability to pay
   vi. willingness to pay.

D. *Specific emission allocation proposals* including:
   i. equal per-capita entitlements
   ii. national historical per-capita emissions
   iii. status quo
   iv. mixed and ad-hoc proposals.

E. *Distributing abatement costs* including:
   i. transfers determined by emissions allocation then tradeable entitlements (carbon trading) with an excess to developing countries
   ii. direct payment accountability approaches
   iii. polluter pays
   iv. historical responsibility
   v. incorporating ability to pay.

F. The concept of joint implementation with richer countries investing in projects in poorer countries to reduce or sequester greenhouse emissions.

Source (adapted from): IPCC, *Climate Change 1995: Economic and social dimensions of climate change.* Contributions of Working Group III to the Second Intergovernmental Panel on Climate Change (Cambridge: Cambridge University Press, 1996).

In attempting to address these, the German government set up a working group to discuss various principles or criteria for the fair allocation of national targets. These included: carbon dioxide emissions per capita (to reflect differing national population sizes and variation of outputs from individuals within populations); level of wealth as Gross Domestic Product (GDP) per capita (to reflect that richer countries could and should bear a greater proportion of costs); emissions intensity of output as carbon dioxide emissions per GDP (to reflect differential contribution to the problem as well as national wealth); national climate characteristics as heating degree days (to take account of differential impacts); dependence on primary energy; and dependence on fossil fuels.

Economic models and their pitfalls

Such welcome efforts to introduce elements of fairness, however, need to be considered against the backdrop of the domination of the economic *Weltanschauung* in which they are placed and indeed calculated. Early on, economic models – collectively termed Integrated Assessment Models (IAMs) – were developed to allow global analysis of greenhouse gases (full cycle of anthropogenic gases; concentrations of gases in the atmosphere), resultant climate change, impacts on the environment and economy (economic losses as the 'damage function'), and the costs of slowing climate change ('cost function'). The belief was that, through 'creating the same metric for cost and benefit assessments, IAMs [could] be used to develop economically efficient policies' [21].

But there is little attention paid in such models to equity concerns and, despite more creative recent inclusions, to some the debate over climate-change policies represents the discomforting conceptual imperialism of economics. As Harvey stresses, 'some sort of hegemonic economistic–engineering discourse has come to prevail which . . . has the effect of making us 'puppets of the institutional and imaginary worlds we inhabit' . . . by commodifying everything and subjecting almost all transactions . . . to the singular logic of commercial profitability and the cost–benefit calculus' [22].

Economic analyses such as IAMs have embedded within their methodology a picture of human manipulation of the natural world, and the natural world seen largely in terms of monetary value. As well as the deep-seated flaws in representing the world in such a way, such analyses also misrepresent that which people actually value [23]:

> Many valuable goods escape the net of the national income accounts and might affect the calculations of the economic effects of climate change . . . Among the areas of importance are human health, biological diversity, amenity values of everyday life and leisure, and environmental quality. Some people will place a high moral, aesthetic or environmental value on preventing climate change, but I know of no serious estimates of what people are willing to pay to stop greenhouse warming.

Yet despite acute criticisms of this kind, the range of criteria brought into the calculations do indicate efforts to address concerns about global social justice within the climate-change debate. Linnerooth-Bayer suggests that Rawls has been instrumental, postulating that 'moral reciprocity in the veil of ignorance forces individuals to treat others as they would want to be treated themselves, making responsibility to fellow humans an intricate functional property of Rawls's justice scheme' [24]. But this assertion perhaps misses the reality. There may have been a shift in social conscience, if not in practical policy outcomes, with Rawls a useful resource to draw on. But this has remained predominantly within an economic framework (and all the limitations embedded within that) built on administrative utilitarian thinking.

Intellectual developments are, however, contained by the value systems in which they occur; they are also only given practical import by resultant action. So it is pertinent to look now at how the theoretical changes have been mirrored in two parallel sets of processes in the climate-change debate: developments in international policy around managing climate change; and growth in the campaigning efforts of pressure groups.

Climate-change policy and climate justice

As is widely known, commitment to policy change has been hard to secure but there has been some progress. In 1995 the Berlin Mandate of the Framework Convention on Climate Change (FCCC)[2] set in train processes that would lead to the mandatory emission reduction targets that would be agreed in Kyoto, Japan, in late 1997. The FCCC contained the following important principles:

- an objective to stabilise greenhouse gas emissions to a level that would prevent dangerous interference with the climate-change system and allow sustainable economic development;
- developed countries to take the lead role in policies to control emissions (goal accepted, but not the target of stabilising emissions at 1990 levels by 2000);
- obligation on richer countries to assist developing countries vulnerable to the adverse effects of climate change in their adaptation strategies (economies in transition not obligated);
- developed countries and other developed parties of Annex II[3] to transfer technology and finances to enable developing countries to implement their more limited commitments;
- economic needs and special circumstances of poorer countries (and those highly dependent on fossil fuels) to be taken account of in determining emissions targets;
- parties to the FCCC to promote sustainable development and take precautionary measures to minimise costs of greenhouse uncertainties and risks [25].

At the 1997 Kyoto conference the European Union (EU) then advocated a uniform 15% reduction in greenhouse emissions below 1990 levels for all Annex I (developed) parties; stronger proposals came from the Alliance of Small Island States (AOSIS),[4] supported by the Association of South-East Asian Nations (ASEAN), for Annex I countries to cut emissions by 20% below 1990 levels by 2005. However, following extensive debate and consideration of various fairness criteria, the Kyoto Protocol agreed instead that: Annex II countries would limit greenhouse gas emissions to an average of 5.2% below 1990 levels in the commitment period 2008–12; EU countries would cut emissions by 8%; the USA would cut emissions by 7%; Japan would cut emissions by 6%; and Russia and Ukraine would stabilise to 1990 levels. Australia was controversially granted an 8% increase in emissions,[5] and developing countries were not expected to agree to mandatory targets for emissions. It was the Kyoto Protocol to the UNFCCC that created three cooperative mechanisms: Emissions Trading; Joint Implementation; and the Clean Development Mechanism.[6]

At the 1998 meeting in Bonn, Germany, discussion moved towards rights, and the principle of equal per-capita entitlements (in respect of greenhouse gas emissions) was put forward. This notion of the equal right to pollute, based on equal ownership of the earth's atmosphere, was framed in terms of global fairness or justice. The principle does have strong egalitarian roots but its practical application tips the balance in favour of richer countries who could, through mechanisms such as carbon trading, purchase from poorer countries the right to pollute more than their 'fair share' [26].

In 2001 at the COP-6 (Conference of the Parties – COP – to the UNFCCC) meeting, the 'Bonn Agreement' – on the Buenos Aires Plan of Action to adopt the Kyoto Protocol – included political compromise and consensus on funding, technology transfer, adverse impacts, flexibility mechanisms, sinks and compliance. This led to the more detailed international climate change policy regime that culminated in the signing of the Marrakech Accords at COP-7 in late 2001.[7] Famously, the USA unilaterally withdrew from these negotiations [27]. The Kyoto agreement came into force in 2005, but recent figures from the European Environment Agency suggest most EU countries are unlikely to meet their targets.

Campaigning for justice in climate change

In parallel to policy developments and negotiations (and sometimes providing evidence for them) there has been a groundswell in 'independent' think-tanks, non-profit-making organisations, and other new bodies established to press for fair and generally more aggressive policy targets relating to climate change. A number of these have expressed their opinions and activities in terms of global justice, and their mix of conscience-driven academics and pressure-group campaigners has provided both the intellectual base and the energy needed to drive activities forward. There is the feel of a throwback to the lobbying efforts of the first half of the twentieth century to clean the skies of air pollution, as described in Part II.

The Global Commons Institute (GCI), for instance, was set up in 1990 in London, and has been encouraging awareness of its solution to climate change called *Contraction and Convergence*. Put forward as the suggested international framework for the arrest of greenhouse gas emissions *Contraction and Convergence* argues that economic growth can continue at current ('business as usual') rates only provided large efficiency gains are made and nearly all energy comes from renewable sources [28]. Dramatic reductions in carbon dioxide emissions would ensue, with the possibility of emissions trading between richer and poorer countries. Without this the GCI estimates that by 2060 the (annual) costs of global damage caused by climate change would equal (and then rise above) the economic gains[8] of increasing global production [29].

The Tyndall Centre for Climate Change Research [30] has a project on climate justice, and has produced papers on related areas. For instance, a 2002 publication highlights the different vulnerability of different sectors and groups to the risks posed by climate change, and stresses the importance of sharing risks. Equitability of adaptation will require deep thinking and careful planning, otherwise unequal structures may be reinforced [31]. The vulnerability of the poorest and the significance of building social capital within National Adaptation Programmes of Action (NAPAs) has been pointed out elsewhere [32].

Another group, the cleverly named US-based EcoEquity, is committed to advancing equal rights to global commons resources, in particular the principle mentioned earlier of equal per-capita rights to the atmosphere. Lamenting both US rejection of the Kyoto Protocol and also the Byrd–Hagel resolution,[9] EcoEquity argues that fairness 'cannot and will not mean that the rich go on as before' and that a climate treaty will have to embody a 'fairness that is acceptable in China as well as the United States.' EcoEquity hopes to deepen and clarify the meaning of climate justice through drawing together academics and non-governmental organisations into the global justice movement: 'What will we be doing?' poses the EcoEquity website, then answers, 'Working to bring the many threads now being spun around climate justice together into a stronger web, one that can support a broader political strategy' [33].

There are other individual groups or organisations [27], but a powerful coalition of groups – including CorpWatch, Friends of the Earth International, OilWatch Africa and the World Rainforest Movement – gathered as the International Climate Justice Network at the final preparatory negotiations for the Earth Summit in Bali in June 2002. The coalition developed a set of principles aimed at 'putting a human face' on climate change. The 'Bali Principles of Climate Justice' first list (as a series of 'Whereas') the nature of the problem (caused primarily by the rich; felt disproportionately by small island states, coastal peoples, women, the poor and others; violating human rights) then state 27 core principles of the international movement for Climate Justice. These include, as numbered by the network:

1. Affirming the sacredness of Mother Earth, ecological unity and the interdependence of all species, Climate Justice insists that communities have the right to be free from climate change, its related impacts and other forms of ecological destruction.

7. Climate Justice calls for the recognition of a principle of ecological debt that industrialised governments and transnational corporations owe the world . . .

8. . . . Climate Justice protects the rights of victims of climate change and associated injustices to receive full compensation, restoration and reparation for loss of land, livelihood and other damages.

17. Climate Justice affirms the need for socio-economic models that safeguard the fundamental rights to clean air . . .

26. Climate Justice requires that we, as individuals and communities, make . . . choices to consume as little of Mother Earth's resources . . . and make the conscious decision to challenge and reprioritise our lifestyles, rethinking our ethics with relation to the environment and Mother Earth [34].

These principles serve well in bringing up to date the prominence of the use of justice and equity in the climate change debate, and also in paving the way for the final sections of this chapter, but it is worth synthesising at this stage. The most recent conception of air and health – as represented by the approach taken to dealing with climate change – has been placed within a framework of social justice, and has spawned a range of academic, policy and pressure group writings reflecting, generally quite loosely, ideas articulated by John Rawls.

This conception has become a forum for expressing dissatisfaction with the perceived reasons underlying many of the world's ills. It is the arena for discussion and analysis of the impact of industrialisation and of modern Western lifestyles, global poverty, the conceptual imperialism of economics, and of liberal democracy as a political endpoint [35, 36]. It is unclear how much, in policy terms, fairness does end up mattering, and Victor may be right in asserting that 'willingness to pay' remains key in the climate-change debate, as elsewhere. Fairness, he suggests, is used as rhetoric in debate, and only climatic catastrophe is likely to generate moral rethinking [37]. But the rhetoric probably does, nevertheless, mark a sea-change in thinking.

Inequalities in health: social justice in public health

One of the criticisms levelled at the climate-change debate is that it focuses on the bigger global issues at the expense of more local practical matters. There is, however, an example of modern incorporation of social justice into public health, which is more contained and has interesting parallels. That is the debate about inequalities in health, of which the points salient to this book will be outlined.

The ground-breaking *Black Report* of 1980 showed that an individual from the lowest social class is likely to have worse health through his or her life, and die younger, than someone better off [38]. The findings of the *Black Report* were notoriously suppressed but were replicated just under a decade later in a follow-up report, the *Health Divide* [39]. They are now well-established and widely held. As well as strides in understanding the relationship between *absolute* material well-being and poor health, awareness more recently has grown of the importance of *relative* parameters [40, 41]. Though still contentious, relative income has been put forward as having a direct bearing on health levels, within and between countries, regardless of the absolute levels of wealth or income [42, 43]. This has been explained through the idea of general susceptibility, mediated perhaps through community-rooted psychosocial factors, such as social cohesion [44]. Living in a country like the UK, with high income inequalities, may make an individual more likely to be unhealthy compared with someone in Sweden (with its low-income inequalities).

Attention to health inequalities was too politically sensitive in the 1980s and early 1990s,[10] but is now more acceptable [45]. In the UK, an independent enquiry into inequalities in health was undertaken in 1998 [46], followed by inclusion of the topic in governmental policy [47]. This has been augmented recently through specific strategies aimed at reducing health inequalities, which include national targets[11] to reduce gaps in infant mortality and life expectancy between those better off and those worse off [48, 49].

Inequalities in health are essentially descriptive but have tended to be perceived as synonymous with the value-laden notion of *inequities* in health. This was clearly illustrated by a useful observation that appeared in the *British Medical Journal* in 2001 [50] on the use of the expression 'inequalities in health' in, for example, the UK, as referring to *inequitable* inequalities (a subclass of the class):

Inequalities in health, formally defined refer to a broad range of differences in both health experience and health status between countries, regions, and socio-economic groups. Most inequalities are *not* biologically inevitable but reflect population differences in circumstances and behaviour that are in the broadest sense socially determined. However, in industrialised countries such as the United Kingdom, the term 'inequalities in health' has tended to refer to differences in health status between regions and population subgroups *that are regarded as inequitable* [author's italics].

The idea underlying the inequalities in health debate is that there is something inherently unfair and wrong about the gap between health experiences of different social groups – whether these are different social classes, different ethnic groups, different geographical areas, or men and women.[12] In a speech to the Faculty of Public Health Medicine in 2002 the then Secretary of State for Health proclaimed it unacceptable 'that the opportunity for a long and healthy life today is still linked to social circumstances, childhood poverty, where you live, how much your parents earned, how much you earn yourself, your race and gender.' The health gap between rich and poor, he added, 'offends against all this government stands for: a society based on fairness and justice, in which each citizen gets the opportunity to fulfil the potential of all their talents' [51].

As with Climate Justice, this too is social justice in public health but with more defined boundaries. Accordingly, a fairer picture of overall population health entails a commitment to narrower distribution of health experiences in social groups that make up the population. New health policy in the UK aims to address this, although success is likely to be limited unless accompanied by fiscal policies that reduce income inequalities [52]. The philosophical underpinning echoes Rawls's ideologically, including the stress on maximising the lot of the worst-off while (presumably) not lessening the lot of those better off. It is also an administrative utilitarian approach, just like social justice in the climate-change debate, contracting the distribution of assets within an essentially closed system, or at least a system that has limits at any point in time although those limits might change with time.

With this in mind it is worth recapping that, so far, the moral foundations of public health philosophy that have been looked at have drawn predominantly on either utilitarianism or, more recently, social justice. The next chapter explores how the ethical foundations can be thrown wide open, as exemplified by environmental ethics. But before that, it is important to recognise that there are other moral frameworks, notably what is offered by virtue ethics and also the perspectives of Wittgenstein.

Other moral frameworks: virtue ethics and Wittgenstein

For several hundred years Western moral philosophy has been dominated by consequentialist and deontological theories, in particular utilitarianism and Kantianism.[13] In the sphere of public welfare, utilitarianism has been challenged in modern times by theories of social justice, as has been discussed. Within clinical medicine and biomedical ethics[14] much attention has been given to duties and rights.[15] Public health straddles medicine and public policy, making relevant the moral traditions key to both these spheres. Before examining a radical challenge to the centrality of these traditions to public health, it is important to consider other directions of moral philosophy. Two of these, in particular, provide a reflective adjunct to the discussions around the final conception of air and health.

Virtue ethics

The idea that guidance on how to live a good life is linked to traits of the character originated with the Greeks, notably with Aristotle, and in the thirteenth century Thomas Aquinas synthesised Aristotelian ethics with Christian theology. Virtue ethics has never entirely ceased to be an important mode of ethical thought, and is built into Western common-sense morality, but particular interest resurfaced in the second half of the last century, especially following an influential article published in 1958 by the respected British philosopher Elizabeth Anscombe [53].

In that article she contended that modern moral philosophy was directionless and inappropriate, and justified her position in part through placing developments in moral philosophy in historical context. Anscombe, and others to follow, argued that the general decline in intellectual activity in the West following the demise of the Greek and Roman empires saw stagnation in moral philosophy through the early medieval period, or so-called Dark Ages (AD 500–1000). When interest revived in the later Middle Ages, the secular approach of the Greeks had been replaced by philosophy drawing heavily on Divine Law, largely in the Judeo-Christian tradition.

This change had two connected effects. First, deference to a creating, supernatural force slanted the focus of moral philosophy towards areas of personal and social life within the religious realm, such as family life; any moral uncertainty could usually be cleared up by reference to a final arbiter with whom it was difficult to contest. Second, the shift directed moral philosophy firmly towards acts, and away from analysis of what a good life might consist of and how one ought to live it. This is what had previously preoccupied Socrates, Plato and Aristotle, but had become of less immediate importance than moral adjudication of the rightness or wrongness of specific actions.

As moral philosophy progressed through the Renaissance and Enlightenment periods the significance of Divine Law waned, especially with the impact of Cartesianism,[16] but the emphasis on acts remained. It is in this context that the two major theories in contemporary moral philosophy emerged – both utilitarianism and Kantianism are secular and essentially oriented towards moral adjudication of actions. As has been already mentioned, the former centres on acts (or rules) maximising well-being, the latter on the role of reason and the accordance of actions with duties. They are both what can be called 'normative' theories in that they provide a set 'of general principles which provide a *decision procedure* for all questions about how to act morally' [54].

In secular modern Western societies, this emphasis on acts makes for a legalistic type of basis for ethics, which loses its sense and purpose without the religious element from which it is derived. Modern moral philosophy can thus appear empty and detached from the human condition, from what makes our lives worth living. As the philosopher Roger Crisp recently reminded us, Anscombe 'believed that a return to an Aristotelian view of ethics, in which norms are founded not legalistically but on a conception of human flourishing with virtue at its centre, provided for moral philosophy the only hope of its retaining any significance' [55].

There are different conceptions of virtue ethics but Aristotle's is perhaps best known. The translations of many Greek words and phrases are problematic but Aristotle believed that living virtuously is inextricably linked to (the goal of) human flourishing, or happiness. Virtues are the subject matter of ethics, since a good man (who leads a good life) is one of virtuous character, one who lives according to virtues. According to Aristotle a virtue is a trait of character displayed habitually out of principle. Generosity, for instance, would be displayed automatically by a virtuous person and not just on occasion. Though some character traits are skills particular to specific jobs or roles in life, the virtues are those common to all persons. One helpful way of assessing what might be a virtue involves thinking about situations when, in trying to decide what to do, one contemplates what a particular revered person (perhaps a mentor or rabbi) might do: it is likely that he or she would be displaying a virtue [56].

Aristotle considered the nature of particular virtues at some length, including many that still feel applicable today: courage; temperance; honesty; compassion; dependability; loyalty; friendliness; moderation; and tolerance. According to Aristotle, each virtue is a mean between two particular vices. For instance, courage or bravery is a virtue lying in the mean between the (deficiency) vice of cowardice and the (excess) vice of foolhardiness. A virtuous man[17] uses reason to decide the appropriate response for any particular situation; he knows for each situation the degree of courage that makes the behaviour appropriately brave, rather than inappropriately reckless or cowardly. The concept of virtue lying in the mean between two vices is therefore not about moderation but about a shifting awareness of the correctly virtuous response – sometimes it will be necessary to act with great courage, sometimes not. Reason is thus inseparably linked to virtue and ethics.[18]

One particular virtue needs to be mentioned here. During the course of this chapter, justice has been considered in some detail. However, though Aristotle did indeed consider justice at length, his attempt to explain it as a virtue was muddled. This is largely because Aristotle's perception of justice was essentially procedural, evident in his subdivisions of rectificatory and distributive justice – the former legalistic and compensatory, and the latter about fairness of spread. Neither correlates directly to a single virtue, and Aristotle may have been more successful focusing on righteousness. As Urmson [57] puts it:

It is not difficult to see what has gone wrong. Distributive and rectificatory justice, both of which can be manifested only by someone who is acting in a judicial or quasi-judicial capacity, simply are not excellences of character at all, nor are they manifestations of a single excellence of character. There is no special emotion that a judge ought to feel and exhibit to a right degree. Rather, he has to operate the principles of justice correctly, and to this end he has to be impartial . . . even-tempered and, of course, clear-headed; in a word, he needs to be a man of generally good character, and not to display some special trait of character. The same is true of the schoolmaster, the government official . . . Administering justice is a special occupation, not a special character trait.

The last point that needs to be made about virtue ethics relates to a current proponent who, like Anscombe, has challenged the domination of contemporary philosophy by Kantianism and utilitarianism; but he has also taken on liberalism. In his 1981 book *After Virtue* Alisdair MacIntyre [58] similarly argues that modern moral philosophy is essentially misplaced. The preoccupation with analytical work and normative theory has accompanied historical developments, during which a key component has been lost – a common conception of the good.

Without any common conception, suggests MacIntyre, adjudication on competing moral arguments is impossible. Rational assessment within any particular argument is of course valid (indeed this has been the analytical focus), but rival arguments are often conceptually incommensurable, operating within different frameworks. Without any common conception of the good, not only can arguments not be compared but each reduces to assertion or preference, linked to emotions provoked.

The proposed alternative resembles a return to Aristotelian virtues. According to MacIntyre, individual fulfilment and flourishing can be attained through the expression of internal goods in the course of 'practice': he defines a practice as a socially established cooperative activity in which internal goods are realised through trying to achieve standards of excellence that are, in part, definitive of that activity. Examples of practice include sports, farming, and the work of academics and artists.

Liberal institutions are, however, hostile to the virtues because they espouse instead dispositions such as obedience, individualism and acquisitiveness. In addition, liberalism is essentially grounded in capitalism, the market economic foundation of which is driven by efficiency and monetary gain. For most individuals this impacts through a working life devoid – or strongly deficient – in practice, in activity that gives life meaning; and, though practice can be engaged in outside work, there is less leisure time available to do so [59]. The corruption of modern moral philosophy is thus associated with the corrupt political and economic organisation of Western societies, and the soulless lives created for the masses.

Virtue ethics presents some kind of purposeful way forward for MacIntyre. The domination of Western moral thought with, for example, utilitarianism is reflected in the problems of global warming. Ideas of social justice may provide a dimension of fairness, but developments will be restricted because of the constraints of liberalism and the capitalist framework within which it is embedded. For the policy-maker, and public health practitioner, virtue ethics may offer a different kind of template with which to approach problems such as climate change.

Wittgenstein

In most moral philosophy textbooks the Austrian-born philosopher Ludwig Wittgenstein hardly gets a mention. Despite being one of the most revered and influential philosophers of the twentieth century, little attention is generally given to what Wittgenstein had to say about ethics. This is, in part, because Wittgenstein had relatively little to say about ethics (certainly compared to his contribution to philosophy of language and logic) but it is also because he felt that the last person one should turn to about ethics is a philosopher.

Wittgenstein's objection to moral philosophy had a different basis to the objections described in the previous section: it was, to a degree, part of a general problem he had with philosophy. Wittgenstein felt that language was the basis of so-called philosophical problems, and focused his work on demonstrating that at the core of such problems were mainly disputes about terms and their meanings. All debate (including his own) is constrained by the limits of language, and language itself is a constantly evolving social instrument, serving many purposes and interacting with other aspects of societal life. Philosophers should, so Wittgenstein thought, concentrate on illuminating linguistic misunderstandings rather than the favoured analytical problem-solving; and the seductiveness of all-encompassing theories should be resisted. Otherwise philosophy was poor use of time, which Wittgenstein certainly felt himself, abandoning academic life for many years to become a school teacher [60].

Moral philosophy, however, was held in additional disdain by Wittgenstein, especially later in his life, for reasons separate to the general linguistic issues outlined above. He believed ethics to be a deeply personal matter, about private conduct on the one hand, but also about the state of one's soul, and what makes life worth living. Of utmost importance was how an individual lived, in terms of inner life and also interpersonal relationships. Honesty, commitment and duty were central, and Wittgenstein was renowned for his sincerity and integrity. He could not understand how one individual could possibly tell another how to live, because one person simply cannot understand the full meaning of another's life.

The direction of moral philosophy in the first half of the twentieth century was therefore extremely troubling to Wittgenstein. The emphasis on complex analytical work within normative moral theories was mistaken because good philosophy should focus on the meaning of the language employed instead. But it was also misplaced because ethics is essentially personal, and not about systematic theories to guide human action. As Wittgenstein eloquently put it 'If I needed a theory in order to explain to another the essence of the ethical, the ethical would have no value at all' [61].

Wittgenstein did not think highly of professional philosophy, and professional moral philosophy in particular – unless it was 'good' philosophy, practised in the way he thought made sense. As Johnston comments, the 'professional moral thinker risks seeking academic success or opportunities to demonstrate the power of her intellect when the real issue is trying to find the right way to live' [62]. Wittgenstein did eventually return to academic life at Cambridge, to continue to challenge modern philosophy, in search of the truth. But his grave misgivings about the professionalisation of philosophy continued; to others he appeared seldom happy, although Wittgenstein remarked towards his death that he felt he had had a wonderful life [63].

It is hard to believe that Wittgenstein would have had much time for the contemporary field of biomedical ethics. Not only is it a subspeciality of moral philosophy, populated by the so-called experts for whom Wittgenstein had little praise, but bioethics is concerned with areas of life deeply personal and human: living; caring; and dying. It is also hard to know what Wittgenstein would have thought about ethical issues in public health, except to emphasise the importance of language, description and representation. But, like MacIntyre, Wittgenstein strongly reminds us that there are alternatives to the bent of modern moral philosophy. These involve seeing and understanding ethics rather differently and, in so doing, reorienting the discussions around issues such as the approach to dealing with climate change.

Other European continental philosophers have also attempted reframing the debate on ethics, notably Nietzsche and Sartre, although in quite different ways.

This chapter has digressed importantly. To bring the pieces together, it is worth restating that the chapter has taken the approach to dealing with climate change as the fourth conception of air and health. The debate about climate change has come to represent social justice as a challenge to utilitarianism, with global health issues at the heart. Within public health, on a more contained national level in the UK, the progression of policy around inequalities in health mirrors justice-oriented moves.

Social justice, however, is not the only challenge to utilitarianism, and indeed seeing it as such misses important points. Renewed interest in virtue ethics displays a discontent with historical trends in moral philosophy,[19] and highlights the connection of these changes to Western political and economic development, and the impact of these changes on personal morality, on living a good life. In the first half of the twentieth century Ludwig Wittgenstein argued that the direction of modern philosophy was deeply misplaced, with additional significant reservations about moral philosophy.

Tying in concerns articulated by advocates of virtue ethics and by Wittgenstein, the next chapter returns the emphasis to the environment and health, and presents a radical challenge to philosophical debate within public health.

NOTES

1. I am using man here, rather than person, to represent Aristotle's depiction, which focused predominantly on men.
2. The United Nations Framework Convention on Climate Change (UNFCCC) aims at the stabilisation of carbon dioxide concentrations at a level that will prevent dangerous anthropogenic interference with the climate system, within a time-frame sufficient for allowing ecosystems to adapt naturally to climate change, to ensure that food production is not threatened and to enable economic development to proceed in a sustainable manner. This multilateral treaty was opened for signature at the Rio Earth Summit in 1992, and came into force in 1994.
3. European Union (EU) and member countries of the Organisation of Economic Cooperation and Development (OECD).
4. These islands are likely to be most directly, and most seriously, affected by changes in climatic conditions (i.e. most vulnerable), and are also in a poor position to adapt or cope with the consequences.
5. Based on the argument presented around its dependence on fossil fuel for domestic energy and export revenue.

6. The Clean Development Mechanism allows developed countries to use certified emissions reduction credits in developing countries, and also assists developing countries in achieving sustainable development targets.

7. The declaration included an adaptation fund for developing countries, a special climate change fund from Annex I country contributions, a least-developed country fund, a technology transfer expert group, limited banking of excess emissions and a compliance committee. Although marking progress and containing important elements, these developments were generally felt not to have been overly satisfactory for developing countries.

8. Economic gains are measured as the total global product, the sum of global domestic products.

9. A campaign prior to the Kyoto negotiations of 1997 led to 95 US senators demanding that developing countries also take on firm reduction commitments, so challenging the UNFCCC principle that developed countries take the lead in reducing emissions.

10. The term 'variations in health' was put forward instead.

11. The targets are, starting with children under one year, by 2010 to reduce the gap in mortality by at least 10% between 'routine and manual groups' and the population as a whole; and starting with local authorities, by 2010 to reduce the gap by at least 10% between the fifth of areas with the lowest life expectancy at birth and the population as a whole.

12. This holds for inequalities at different levels: determinants of health (environment, lifestyles); health experiences (mortality, morbidity); and access to health services. Much attention has indeed been on this latter aspect, equity in health service provision.

13. Deontological theories are those that are duty-based, the moral rectitude of actions judged by accordance with certain duties. The most famous proponent of this is the German philosopher Immanuel Kant (1724–1804).

14. I am using the term 'biomedical ethics' to cover a field of enquiry, to which several similar phrases have been applied, for example bioethics, medical ethics and healthcare philosophy.

15. Some criticisms of developments in medical ethics, which are relevant to this book, are looked at in the conclusions.

16. Cartesianism is considered further in Chapter 11, including its relevance to developments in political philosophy, science and thinking about the environment.

17. Women, and those down the social order, were generally not considered by Aristotle.

18. Aristotle associated each virtue with an emotion. Disposition of appropriate emotion, determined by reason, is excellence of character.

19. The dominant Western approach to biomedical ethics has been similarly challenged by alternatives such as virtue ethics, feminist ethics, narrative ethics and sociological perspectives. This is touched on further in the conclusions of the book.

REFERENCES

1. Quoted in J. O. Urmson, *Aristotle's Ethics* (Oxford: Blackwell, 1998), p. 76.
2. R. Gillon, Four principles plus attention to scope. *BMJ*, **309** (1994), 184–5.

3. T. L. Beauchamp and J. C. Childress, *Principles of Biomedical Ethics* (New York: Oxford University Press, 1994).

4. N. Daniels, Accountability for reasonableness. *BMJ*, **321** (2000), 1300–1.

5. J. Rawls, *A Theory of Justice* (Oxford: Oxford University Press, 1999).

6. *Ibid.*, p. 3.

7. *Ibid.*, p. 5.

8. *Ibid.*, p. 10.

9. *Ibid.*, p. 11.

10. *Ibid.*, p. 118.

11. *Ibid.*, pp. 53, 72.

12. *Ibid.*, p. 86.

13. *Ibid.*, p. 133.

14. *Ibid.*, p. 68.

15. S. Rayner, E. L. Malone and M. Thompson, Equity issues and integrated assessment. In F. L. Toth, ed., *Fair Weather: Equity concerns in climate change* (London: Earthscan, 1999), pp. 11–43.

16. C. Hamilton, Justice, the market and climate change. In N. Low, ed., *Global Ethics and Environment* (London: Routledge, 1999), pp. 90–105.

17. J. E. Reichart, 'The tragedy of the commons' revisited: a game theoretic analysis of consumption. In L. Westra and P. H. Werhane, eds., *The Business of Consumption: Environmental ethics and the global economy* (Oxford: Rowman & Littlefield, 1998), pp. 47–66.

18. M. Ha–Duong, M. J. Grubb and J.-C. Hourcade, Influence of socioeconomic inertia and uncertainty on optimal CO_2-emission abatement. *Nature*, **390** (1997), 270–3.

19. M. Hayry, European values in bioethics: why, what and how to be used? *Theoret. Med. Bioethics*, **24**:3 (2003), 199–214.

20. IPCC, *Climate Change 1995: Economic and social dimensions of climate change.* Contributions of Working Group III to the Second Intergovernmental Panel on Climate Change (Cambridge: Cambridge University Press, 1996), p. 104.

21. F. L. Toth, Fairness concerns and climate change. In F. L. Toth, ed., *Fair Weather: Equity concerns in climate change* (London: Earthscan, 1999), pp. 1–10.

22. D. Harvey, Considerations on the environment of justice. In N. Low, ed., *Global Ethics and Environment* (London: Routledge, 1999), pp. 116–17.

23. C. Hamilton, Justice, the market and climate change. In N. Low, ed., *Global Ethics and Environment* (London: Routledge, 1999), p. 101.

24. J. Linnerooth-Bayer, Climate change and multiple views of fairness. In F. L. Toth, ed., *Fair Weather: Equity concerns in climate change* (London: Earthscan, 1999), p. 54.

25. United Nations, *United Nations Framework Convention on Climate Change.* http://unfccc.int (accessed 17 May 2005).

26. B. Mueller, *Equity in Climate Change: The great divide.* www.oxfordenergy/pdfs/great_divide_executive_summary.pdf (accessed 17 May 2005).

27. Teta Energy Research Institute, *Climate Change.* www.teriin.org/climate/climate.htm (accessed 17 May 2005).

28. A. Meyer, *Contraction and Convergence: The global solution to climate change* (Dartington: Green Books, 2000).

29. Global Commons Institute, *Basic Climate Scenarios.* www.gci.org.uk/scenarios.html (accessed 17 May 2005).

30. www.tyndall.ac.uk.

31. W. N. Adger, S. Huq, K. Brown, D. Conway and M. Hulme, *Adaptation to Climate Change: Setting the agenda for development policy and research* (Norwich: Tyndall Centre for Climate Change Research, 2002).

32. S. Huq, A. Rahman, M. Konate, Y. Sokona and H. Reid, *Mainstreaming Adaptation to Climate Change in Least Developed Countries (LDCs)* (London: International Institute for Environment and Development, 2003).

33. EcoEquity, *About EcoEquity.* www.ecoequity.org/about.html (accessed 17 May 2005).

34. CorpWatch, *Bali Principles of Climate Change.* www.corpwatch.org/article.php?id=378 (accessed 17 May 2005).

35. T. Athanasiou and P. Baer, *Dead Heat: Global justice and global warming* (New York: Seven Stories Press, 2002).

36. D. Brown, *Ethical Problems with the United States' Response to Global Warming* (Blue Ridge Summit: Rowman & Littlefield, 2002).

37. D. G. Victor, The regulation of greenhouse gases: does fairness matter? In F. L. Toth, ed., *Fair Weather: Equity concerns in climate change* (London: Earthscan, 1999, pp. 193–206.

38. Department of Health and Social Security, *Inequalities in Health: Report of a research working group* (London: DHSS, 1980).

39. M. Whitehead, The health divide. In P. Townsend, N. Davidson and M. Whitehead, eds., *Inequalities in Health: The Black Report and the Health Divide* (Harmondsworth: Penguin, 1992), pp. 221–400.

40. Y. Ben-Shlomo, I. R. White and M. Marmot, Does the variation in the socio-economic characteristics of an area affect mortality? *BMJ*, **312** (1996), 1013–14.

41. R. Wilkinson, Socioeconomic determinants of health: health inequalities – relative or absolute standards? *BMJ*, **314** (1997), 591.

42. —, *Unhealthy Societies: The afflictions of inequality* (London: Routledge, 1996).

43. M. Wolfson, G. Kaplan, J. Lynch, N. Ross and E. Backlund, Relation between income inequality and mortality: empirical demonstration. *BMJ*, **319** (1999), 953–7.

44. M. Marmot and R. G. Wilkinson, Psychosocial and material pathways in the relation between income and health: a response to Lynch *et al. BMJ*, **322** (2001), 1233–6.

45. M. Whitehead, *Health Inequalities.* The Official Newsletter of the ESRC Health Variations Programme, **1** (1998), 4–5.

46. Anonymous, *Independent Inquiry Into Inequalities in Health* (London: The Stationery Office, 1998).

47. Department of Health, *The NHS Plan* (London: Department of Health, 2000).

48. —, *Tackling Health Inequalities: Consultation on a plan for delivery* (London: Department of Health, 2001).

49. Department of Health, *Tackling Health Inequalities: Summary of the 2002 cross-cutting review* (London: Department of Health, 2002).

50. D. A. Leon, G. Walt and L. Gilson, International perspectives on health inequalities and policy. *BMJ*, **322** (2001), 591–4.
51. A. Milburn, Speech given to the Faculty of Public Health Medicine, 22 November 2002. Available on Faculty of Public Health Medicine website, www.fphm.org.uk (accessed 15 December 2002).
52. G. Davey-Smith, J. N. Morris and M. Shaw, The independent inquiry into inequalities in health. *BMJ*, **317** (1998), 1465–6.
53. G. E. M. Anscombe, Modern moral philosophy. *Philosophy*, **33** (1958), 1–19.
54. R. Hursthouse, Normative virtue ethics. In R. Crisp, ed., *How Should One Live? Essays on the virtues* (Oxford: Oxford University Press, 1999), p. 31.
55. R. Crisp, Modern moral philosophy and the virtues. In R. Crisp, ed., *How Should One Live? Essays on the virtues* (Oxford: Oxford University Press, 1999), p. 2.
56. G. Pence, Virtue theory. In P. Singer, ed., *A Companion to Ethics* (Oxford: Blackwell, 1997), pp. 249–58.
57. J. O. Urmson, *Aristotle's Ethics* (Oxford: Blackwell, 1998), pp. 76–7.
58. A. MacIntyre, *After Virtue* (London: Duckworth, 2002).
59. A. Mason, MacIntyre on modernity and how it has marginalized the virtues. In R. Crisp, ed., *How Should One Live? Essays on the virtues* (Oxford: Oxford University Press, 1999), pp. 191–210.
60. A. Kenny, *Wittgenstein* (Harmondsworth: Pelican Books, 1975).
61. Quoted in C. Elliott, Introduction: treating bioethics. In C. Elliott, ed., *Slow Cures and Bad Philosophers: Essays on Wittgenstein, medicine, and bioethics* (London: Duke University Press, 2001), p. 3.
62. P. Johnston, Bioethics, wisdom and expertise. In C. Elliott, ed., *Slow Cures and Bad Philosophers: Essays on Wittgenstein, medicine, and bioethics* (London: Duke University Press, 2001), p. 159.
63. K. E. Tranoy Wittgenstein: personality, philosophy, ethics. In C. Elliott, ed., *Slow Cures and Bad Philosophers: Essays on Wittgenstein, medicine, and bioethics* (London: Duke University Press, 2001), pp. 181–92.

11

The bigger picture: environmental ethics and new moral horizons in public health

In Chapter 9 the science of climate change was presented, together with an overview of utilitarianism; critiques of utilitarianism were explored in relation to the climate-change debate, and with reference to public health more generally. John Rawls's theory of social justice was looked at in Chapter 10, as a contemporary alternative ethical framework that has been propounded within the climate justice literature, and also as an idea that features in the debate around inequalities in health. Other challenges to the direction of modern moral philosophy were put forward. Chapter 11 endeavours to pull the pieces together, the pieces of Part IV and also of the whole book. It is argued that developments in moral philosophy, political philosophy, science and public health are connected, and are all relevant to the approach to dealing with climate change.

As a backdrop to Chapter 11, it was a little under 400 years ago that the French philosopher René Descartes established, while tucked alone inside a stove, that the only thing one can truly know, that is beyond doubt, is that one exists: 'cogito ergo sum' or 'I think therefore I am'. Descartes is renowned for his Discourse on Method and Meditations, a set of readings in which he separates mind (where thinking is located) from matter (which can be explained mathematically), dubiously proves the existence of God, and proclaims a long-needed severance with the Greek philosophy that had dominated Western thought for almost two millennia [1]. Discourse remains a key text in the history of philosophy, with Descartes systematising a trend in thinking. But the impact of the ideas within that book – and work by others to follow – could surely not have been anticipated even by the somewhat arrogant author himself.

Cartesian philosophy marked a change in the way we think about and understand the world,[1] altered how we perceive ourselves and relate to others, and shifted beliefs about social organisation and the place of citizens within society. Present-day environmental philosophers and environmentalists have argued fiercely that the roots of current environmental crises lie deep in the seventeenth century, and the importance of attention to roots should never be underestimated – any reorientation of the way we understand the world needs to tackle the foundations.

So the final chapter initially sets out how the period around Descartes changed our view of ourselves, including our moral outlook, shaped the way for modern science and medicine, and had a bearing on Western political philosophy. These developments are not separate but closely interwoven. Next, an outline is traced of perspectives in environmental philosophy on past and current developments. Here we will be able to see how the last conception of air and health – the approach to dealing with climate change – fits in, albeit a little uncomfortably. The links to public health can then be assessed.

Separation and disconnection: science, nature and political philosophy

Prior to the seventeenth century, Western medicine still drew strongly on Greek ideas and beliefs, even though practical aspects had changed somewhat. As was shown in Part I, in broad terms health was aligned with balance in the bodily humours as well as equilibrium with the natural environment. Disease was understood as bodily imbalance, or disturbance of the equilibrium, interpreted by physicians through symptoms and signs. Treatment was geared towards the reorientation of balance, both within the individual and with nature, and remedies used were of natural origin.

Cartesianism paved the way, and allowed for, the development of what we think of now as modern science. In so doing it also directed the progression towards Western scientific medicine, based as it is on the physical sciences. Making the distinction between mind and matter, however, not only corresponded with creating demarcations between values and facts, but it also changed the way in which we think about those values and facts themselves.

In the centuries after Descartes, huge strides were made in understanding the facts about the way the world works, both the physical world that surrounds us and also the matter that makes up our own bodies. Mechanistic philosophy pictured the world as a machine, with explanations needed for the mechanisms hidden behind the phenomena that we see or otherwise come to observe. Mechanisms have causes and effects and, however complex these may be, they can be broken down to simpler mechanisms as a means to understanding the larger processes better. Conflicts between mechanistic philosophy and long-held Greek beliefs about the cosmos constructed according to mathematical order (alongside ideas of perfect worlds) were gradually to be resolved, and what slowly emerged was the belief that mechanisms could be explained through objective scientific truths [2].

The search for these truths went in different but related directions. Understanding the workings of the outside world has been pioneered through physics, Newtonian and experimental, later contracting to that which is often held up as the 'purest' scientific enterprise, particle physics, explaining nature's mechanics through relationships of the smallest entities at the most intimate level. Alongside astronomy, chemistry and, more recently, geophysics, the drive has been towards the belief that understanding what makes the world tick equates with comprehending how the physical world operates, and this can only be attained by objective scientific examination of constitutive elements. Dividing the world up into disciplines of enquiry aids concentration on the different elements, which themselves need breaking down into more manageable parts, because otherwise the world is too complex and messy. As was seen in Part III science requires reducing the area of interest to a level at which relationships can be demonstrated [3].

Developments in the science of the outside world were largely mirrored in scientific understanding of the human body, once this had been accepted as essentially matter, subject to the same laws of physics that operated elsewhere. William Harvey's rejection of Galen marked a turning-point in understanding human circulation and blood, with parallel progress in human anatomy and physiology, and a growing appreciation of the physical and chemical (later biochemical) processes that make the human body work. In Part I we saw how understanding of the biology of disease processes progressed steadily during the eighteenth and nineteenth centuries, although it was the turn of the twentieth century that heralded the significant breakthrough around the pathological basis of infectious diseases.

As with the outside world, division of the body by fields of study has proven a necessity of scientific explanation, but in more recent times the body has been further compartmentalised. As more and more detailed examination of the functioning of the body – through, for example, cell biology, cell pathology, electron microscopy and genetics – has been accepted as most likely to yield important information, medicine has had to specialise to accommodate or foster these pathways to knowledge. Apart from the community-based general practitioner (family doctor), the general physician in a Western hospital is more or less a thing of the past, replaced by cardiologists, nephrologists, hepatologists, urologists and brain surgeons, each now with their own subspeciality or focused area of special interest. But, as will be returned to shortly, in an effort to make the complex more simple, and more amenable to scientific investigation, relationships change. Not only is it questionable whether what holds at a (sometimes artificially) reduced level also pertains at the higher level from which it was taken, but the process of reduction and simplification alters the way we see the world. Smaller concepts, simplified systems, linear relationships and more direct connections can all seem to become more important than that from which they came.

At this point a pivotal question arises, whether the way we think about our material selves and the material world we live in is connected to the way we think about our duties to other humans, animals and the natural world. A scientific interpretation or response might argue that the two should be unrelated, science only serving to better explain objective truths about the world and how it operates. But this is where the demarcation between fact and value is misperceived. While the need for the two to be separate is a core necessity to the scientific enterprise itself, believing that the world can be split and explained in such a manner entails commitment to a way of seeing the world that is value-laden in itself, a commitment at minimum to the reductionist vision. A scientific worldview is different to others, neither right nor wrong, just different, but the mistake is to elevate the scientific way of seeing the world to some kind of superior status. As the philosopher Mary Midgley eloquently suggests, the maps found at the front of an atlas are all needed to describe the world, and it is inappropriate to translate them into a single one of overarching explanation [4].

Any way of seeing the world necessarily involves a belief about that which should be valued within the world, a moral component. It is this connection which will ultimately help illuminate today's moral commitments in public health. The reductionism of modern science, the progressive attention to (and therefore in time valuing of) the smaller over the whole, can be seen reflected in parallel, connected, developments in moral and political philosophy. After all, the moral responsibilities that we feel are appropriate to other humans ultimately shape the institutions that dictate societal responsibilities. Any given political structure will reflect the duties individuals feel towards fellow citizens [5].

A short tour around political philosophy

The most famous example of Greek political theory is Plato's *Republic*, in which the author sets out what he believes to be the ideal society, or utopia. Although Plato holds that the basis of human action is practical self-interest, and left alone man will inevitably favour bodily self-promoting pursuits over others, the tone of the work is about the organisation of a better society – one which functions well and justly, and one that encourages individuals to think and behave beyond their own self-interest. This cooperative society acknowledges that individuals have different skills and promotes that difference, creating an environment in which pursuit of individuals' proper function is fostered. This might be creating health for the Greek physician, or being victorious in battle for the Greek soldier [6].

The challenge for the rulers, deliberately minimally rewarded philosophers,[2] is to get the balance right, to achieve a state in which human happiness, or flourishing, is maximised. In this sense one can see the buds of utilitarian thinking to come almost 2000 years later, but Greek political philosophy stressed rather differently the importance of individual and social diversity. Allowing individuals to realise their potential would symbiotically create a living environment in which social roles and functions were met. In this sense the Greek State could be pictured as a human body, the good functioning of each part necessary for the proper functioning of the whole. There was a very close connection between individuals leading and realising a good and virtuous life, and the just ordering by rulers that created balance and harmony within the organism as a whole.

Plato's utopian society was conceptualised around the size of a Greek city-state, rather than the larger political and national domains that were to come, but its visionary ideas and principles have proved long-lasting. It is difficult to think of other examples of organised society in which the ethical pursuit of individual fulfilment is so closely woven into a cooperative and supportive venture. One reason could be that the guardians who oversaw progression were rewarded for their activities by intellectual and moral gain, rather than by money and power. The republic that Plato heralded was paternalistic in the sense that decisions were made on behalf of citizens by those in charge, but removal of excess material gain separated unwanted incentives. The hundreds of years to follow in Western countries, through the Middle Ages, witnessed what happened when rulers attained wealth through their governance. Even in more recent examples of dictatorial cooperative societies, such as those built on Communist ideals, the leaders become the rich. Perhaps the closest contemporary comparison to Plato's State is something between the Israeli kibbutz and Castro's Cuba [7].

In sixteenth-century Italy Niccolò Machiavelli certainly trusted citizens more than those in power – State or Church – in terms of encouraging stability and liberty, and his thinking symbolised a move away from political paternalism towards more democratic ideals. A century or so later the English philosopher Thomas Hobbes bridged an important gap between the Greeks, Machiavelli and contemporary thought. Whereas Plato and Aristotle felt that humans were naturally inclined, to a degree (or some of them) at least, towards acting virtuously, striving for excellence and organising themselves into social organisation that would foster this, Hobbes did not.

In *Leviathan*, a seminal work in political philosophy, Hobbes provides a stark account of human nature as driven by self-interest. In Hobbes's brutal world people are like machines, driven by appetites (desires) and aversions and will primarily, and somewhat instinctively, look after themselves. Desires encompass the competitive drive to promote personal interests and gain including kin, reputation, material comfort and fortune; aversions include the opposites of these along with a compulsion for safety. This is the natural situation when human beings are left to themselves, the basis of which is, in modern philosophical language, psychological egoism.[3]

To counter the inevitability of conflict and violence in the left-alone situation, man creates the commonwealth or State (Hobbes's great Leviathan), the role of which is to procure for those contained within it an opportunity to better one's interests, while providing a degree of safety. This is ensured through social organisation, enforced by legal mechanisms and force as necessary, with either one ruler (monarchy), rule by a part of society (aristocracy), or rule by an assembly of all (democracy). Citizens participating in such a relationship are in essence signing a social contract, the ascribing to which confers certain benefits and protection, but allows for harsh punishment of behaviour deviating from the contractual agreement. Solitary man, Hobbes believes, is amoral, and it is society that creates notions of rights, fairness and justice. The right to self-preservation, however, remains above all unnatural interference, and from this supposed fundamental right further human rights have been subsequently deduced or derived [8].

Hobbes's vision is not kind on human beings but some would argue it is realistic. He shares Jean-Jacques Rousseau's later reference to a 'state of nature' but differs in seeing it as harsh and unfriendly, as opposed to the romantic's image of a place stripped of society's pointless, artificial desires (especially property) that drive material competition and ultimately create and fuel inequality [9]. Crucially, however, Hobbes sets out a central connection between individualism, morality and political philosophy. The focus on self-interested individuals, set apart from – but also part of – social organisation that serves them, provided the foundations for the application of Darwin's ideas that were looked at in Chapter 2, and will be returned to shortly.

The emphasis on individuals and their entitlements was taken further towards the end of the seventeenth century by another English philosopher, John Locke. Though Locke disagreed strongly with Hobbes's social-contract concept, believing the arbitrariness of its power-holding made it worse than the natural state it apparently rose above, he too stressed the importance of the rights of individuals. Unlike Hobbes, Locke placed a firm belief in human rationality and conscience informing moral judgement (even in the natural state), with ultimate reference to God, holding that no one ought to harm another's life, health, liberty or possessions. Allowing individuals to act according to their own will is grounded in the moral guidance that rationality provides, and such liberty should only be constrained by the State when it interferes with the freedom of another human being [7].

According to Locke the State, however, has a specific civic role around protection of property. To start with, individuals have the property of their own person, and the right to preservation of this, but property rights extend to include the fruits of one's labour. Predating Marx by a couple of centuries Locke believed the value of property lay in the labour put into its construction but, unlike Marx, he argued that effort put into acquisition of materials granted some kind of ownership right. In a famous example Locke postulated that the labour of collecting acorns makes them the gatherer's, by private right, and without needing the approval of all mankind.

Locke had in mind North America as a colony when he thought of land as being freely available, ownership defensibly resulting from mixing one's labour with the soil. Modern ideas of property rights are connected to Locke, and the important notion of the natural environment (which could include air) as something that can be wantonly, and morally, appropriated by humans will be returned to shortly.

Through his emphasis on personal freedom with minimal State hindrance, private ownership and reward of enterprise, John Locke set the tone for modern political liberalism. His core belief that those without basic means of sustenance have fair call on the surplus goods of others (who should provide the transaction) laid foundations for the modern welfare State, which facilitates such processes, and his use of the language of rights influenced Thomas Paine, whose classic book *The Rights of Man* so informed the American constitution and embedded the vocabulary of human rights in that country's moral voice. The impact of Locke was substantial.

Breaking down

With this overview in mind it is now possible to bring together some of the ideas from earlier in Part IV, as well as earlier in this chapter. What we are now able to see is how the seeds of modern political, moral and scientific thought were planted in the sixteenth and seventeenth centuries, took root through the Enlightenment, and were further developed and advanced during the nineteenth and early twentieth centuries. The origins of today's environmental problems, and the relevance of this to public health, can be traced through the interconnected paths of progress in science and political philosophy over the past few hundred years.

Cartesian dualism split mind from matter and, in so doing, began the separation of facts from values. Removing feelings and experiences from bodily workings laid the material elements open to scientific explanation through investigation. In time, and due to its success in predicting the world and providing means to material improvement and power, science has attained a lofty position. Believing the facts to be demonstrated scientifically to be free from values has been part and parcel of progressing the notion that science itself is a value-free enterprise. It is through this notion, the commitment to – and belief in – its objectivity that science has reached its high status. Although the scientific ideology has been widely questioned in the twentieth century, it holds strongly today that the purer the science, and the more objective, the better.

The search for the objective, however, has meant reducing, and then further reducing, complex natural relationships to far simpler ones. This includes, for example, understanding human, animal and plant biology. But changing the way we choose to examine ourselves and others has changed the way we view the world, and the values we place on different aspects of that world. First, in attempting to explain scientifically how we function, we see ourselves as individual entities disconnected from the natural world, whether that world concerns other humans, animals or plants. Then we take it further, for instance by narrowing the view of ourselves to a particular physiological system, and creating specialities and then subspecialities of that area. But the natural world, it would seem, does not really exist or work like that, and modern thought – even scientific – seems to suggest we are much more connected than we realised.

The Greeks were aware of this, in their understanding of the importance of balance and harmony to human health, both for the individual and in relation to the natural world. The political philosophies of Plato and Aristotle mirrored this perspective in picturing the social organisation of the Greek city-state as an organism, different elements working together for the good, or just, functioning of the whole. Within this framework fitted the individual pursuit of a virtuous life.

However, in the centuries of Hobbes and Locke this outlook changed dramatically. The development of mechanistic philosophy and the progression of science somewhat removed humans from the natural environment, which itself was mechanistically objectified. The era of individualism had begun, with justification of self-interested behaviour and an emphasis on individual and private rights. Personal morality no longer had a special relationship to the State, whose role became that of partner in a dispassionate arrangement that primarily provides an environment suitable for promotion of the individual. Mary Midgley summarises [10]:

Since the Renaissance, this kind of contraction has in any case been happening in political philosophy in the West. Political thinkers of the Enlightenment systematically shrank morality by making it essentially a civic affair – a matter of mutual bargaining between prudent citizens within a limited society. Contract thinking sought to abolish the idea of duties towards anyone or anything outside that society . . . That limitation had originally a most respectable aim. It was meant to debunk supposed duties towards the supernatural because those duties had been used to justify religious wars and oppressions . . . But this move had unintended side-effects. It now makes it quite hard for us to make sense of our responsibility towards humans outside our own society, and almost impossible to explain our responsibilities towards non-human nature.

As Locke's earlier example illustrated the era also proclaimed mankind's dominion over nature. The natural environment was articulated in inert, demarcated terms, largely devoid of value, and humans would be morally justified in manipulating it to further legitimate personal interests. This tied in with ownership, rather than stewardship, of nature, and began to set in stone an image of the natural environment – detached and there for human needs – which, in the main, has only relatively recently been challenged by environmentalists [11].

In fact, despite some romantic inclinations, this image of nature was reinforced during the eighteenth and nineteenth centuries as utilitarian political philosophy took hold. As discussed earlier in Part IV, utilitarianism has (indirectly) reinforced moral justification for individual pursuit of that which gives pleasure, with max-imising human happiness as the overall goal. Manipulating nature to meet these ends has ethical validation, and modern welfare economics – the corner-stone of liberal democracies – is grounded in these ideals. Yet utilitarianism, as illustrated, focuses proximally, both in terms of the 'audience' within its calculation (failure to include impacts on those at a distance) and with regard to time – the difficulty of incorporating the needs and desires of future generations, and there is little or no accounting for the intrinsic worth of nature.

As shown in Chapter 2, utilitarianism emerged politically at a time of corre-sponding changes in science, medicine and biology. The connection of human health with nature through miasmatic theories of disease was gradually replaced at the end of the nineteenth century by bacteriological explanations, which catalysed the reductionism of medical science. At that time Darwin and his colleagues were providing a vision of nature that placed self-interested behaviour at its very core, the driver for change, integrally related to adaptation to, and manipulation of, the environment. Not only did this vision reinforce utilitarian thinking, it also provided a basis for ideas of hierarchies of human social organisation, and justification of social Darwinism. The far-reaching impact cannot be underestimated, allowing for articulation of very modern theses that place genetic self-interest at the heart of human behaviour, as espoused for instance in 1976 by Richard Dawkins (and by him and others since) in his popular classic *The Selfish Gene* [12].

Brought together, developments over the last 400 years or so have – in secular Western living at least – disconnected mankind from nature through a mixture of mechanistic philosophical understanding, scientific and medical reductionism, validation of self-interested behaviour and utilitarian-based political philosophy. Individuals have simultaneously become seen as disconnected from other indi-viduals, flitting around as distinct particles within some form of social apparatus, separated from others and the natural world, with purpose, values and goals nar-rowly defined. Midgley again captures this well [13]:

It is the *social atomism that lies at the heart of individualism* – the idea that human beings are essentially separate items who only come together for contingent reasons of convenience. This is the idea expressed by saying that the state is a logical construction out of its members, or that really there is no such thing as society. A social contract based on calculations of self-interest is then supposed to account for the strange fact that such things as human societies do actually exist.

This impacts on all our thinking and actions, public health being but one example. There has, however, been a recent backlash in the form of environmental ethics, which has provided a different way of understanding the world. A look at this philosophical perspective will enable us to comprehend better the conception of air and health examined in Part IV, and to synthesise the implications for public health thought and behaviour.

Environmental ethics

Although environmental ethics has blossomed as an academic activity over the last two decades, its main tenets can be traced back to earlier in the twentieth century. Although these fundamentals have been subject to considerable theoretical and philosophical debate, they have also become inescapably linked to sociopolitical ideologies and movements [14].

It is difficult to place the various philosophical perspectives on the environment into a bag labelled 'Environmental ethics' as they differ in many important aspects, but what they share is a fundamental questioning of the value or values ascribed to nature. Yet even here there are different approaches, or ways in, to examining this core. One such approach, a sort of starting point in environmental ethics, is to distinguish between anthropocentric (human-centred) and non-anthropocentric ethics. This is a good place to begin because an often-shared belief in environmental ethics is that the roots of today's environmental problems lie in the moral favouritism given to human interests, this in itself linked to developments in science and political philosophy discussed earlier in this chapter. The anthropocentric moral favouritism is then disapproved of in different ways, and for different reasons.

In trying to summarily address what an environmental ethic is, Robert Elliot captures this overview, and presents five subdivisions. The first, 'human-centred ethics', has modern utilitarianism as an exemplar, in which facts are needed to calculate the happiness yielded by options, but only humans are treated as morally considerable, i.e. only humans are included in the calculus. An 'animal-centred ethic' treats individual animals as morally considerable but may allow ranking to account for different interests and capacities. Treating equal interests equally and unequal ones unequally, for example, would accommodate ranking humans above animals based on a different capacity for rational autonomous action. A 'life-centred ethic', on the other hand, counts *all* living things as morally considerable, not just humans or non-human animals. However, while some would ascribe equal moral considerability to all life, such as the 'biotic egalitarianism' of Norwegian philosopher Arne Naess, others allow differentiation, for instance by complexity. This may favour, for example, the biosphere over humans, and leads to a special kind of life-centred ethic termed 'ecological holism', which grants moral considerability to wholes, such as large ecosystems or the biosphere: individuals or species are only important in relation to these wholes. The final environmental ethic, called 'rights for rocks' by Elliot, extends moral considerability to all as an 'everything ethic' [15].

Underlying these divisions, or different perspectives, is the justification for any kind of environmental ethic, which is the question of what makes something worthy of moral considerability – worthy of consideration when judging the morality of action [16]. Humans are morally considerable because they have interests that can be promoted or harmed, based on their human capacities – for rational thought and action, and sentience. However, not only is sentience shared by some animals (which could extend moral consideration to them), but moral considerability could lie elsewhere, in some other intrinsically valuable property, for instance complexity or even beauty. This in turn would shift moral considerability to non-sentient animals, plants,[4] ecosystems or the wilderness, and could include non-living entities.[5]

The different perspectives within environmental ethics lie within a spectrum, which stretches from humans to animals, plants, all living and non-living things, incorporating different concepts of what matters morally. Des Jardins, for example, divides the spectrum up a little differently but it still incorporates the same elements. His grouping are: biocentric ethics, which is centred around (all) life and has cor-relative duties;[6] ecological ethics, which focuses on ecological communities and embraces ethical holism; the 'land ethic', articulated first by Aldo Leopold in 1949, which embraces living things, ecosystems and the land [17]; 'deep ecology', especially that of Naess, which emphasises the deep roots of environmental crises, the radical cure needed (personal and cultural transformation) and forcibly expresses its distinction from shallow anthropocentric environmentalism; and social ecology and ecofeminism, which explore how social structures serve the interests and power of certain groups, reflected in and reinforced by domination over nature [18].

However the continuum within environmental ethics is separated out academically or theoretically, a common thread is the difficulty, or failure, to ascribe 'inherent' value to non-human nature, whether that be other animals, vegetation or alternative concepts of what might exist. The anthropocentric nature of Western ethics gives, at best, instrumental value to anything non-human; in other words wombats, wild flowers and the wilderness are of value only by way of serving human interests – as pets, for rambling or as potential new medicines. This has arisen because of entwined developments in science, medicine, and moral and political philosophy that have already been described. It may be fair to reflect that dominion over beasts was heralded back in Aristotelian times but the contemporary situation is rather different in terms of the success and value placed on liberal individualism, materialism, and the sociopolitical structures enshrining these ideologies. The present situation is also vastly different in terms of the depth of environmental crises affecting the planet, of which greenhouse warming is just one example. The Australian philosopher Peter Singer, for instance, despite holding sentience alone as morally considerable [19], is sure of the seriousness of the problem, and the extent of change needed:

Now we face a new threat to our survival. The proliferation of human beings, coupled with the by-products of economic growth, is just as capable as the old threats of wiping out our society – and every other society as well. No ethic has yet developed to cope with this threat. Some ethical principles we do have are exactly the opposite of what we need. The problem is that . . . ethical principles change slowly and the time we have left to develop a new environmental ethic is short. Such an ethic would regard every action that is harmful to the environment as ethically dubious, and those that are unnecessarily harmful as plainly wrong.

Singer then outlines his environmental ethic as including consideration of all sentient creatures now and well into the future, aesthetic appreciation of wild places and nature, rejection of materialistic ideals, promotion of frugality and reassessment of extravagance. He espouses these further, and in more detail, in his acclaimed book *How Are We To Live? Ethics in an age of self-interest* [20]. For Singer and many others the connection between environmental ethics and environmental activism (or environmentalism) is strong.

Putting it all together

It is now possible to start bringing together the different threads of this chapter, and to unravel what is embodied in our final conception of air and health, as represented by the approach to dealing with climate change. Through the 1990s ideas and values articulated within environmental ethics have spawned developments in two related directions: global ethics and environmental justice. Global ethics, and a spin-off, global bioethics, are both broadly concerned with relationships between current Western values, damage to the natural environment (often in global terms, for instance around acid rain or loss of biodiversity) and impacts on human health. The latter with its 'bio-' prefix pays special attention to the place of the healthcare system in the same debates.

Alongside has been the emergence of environmental justice. With a broad agenda, this field of academic debate and social activism has been concerned with many of the same areas as global ethics, but with special attention to fairness of distribution of environmental burdens and benefits, as well as just treatment of individuals in respect of environmental matters. Issues of interest range from unfair distribution of the causes of ozone depletion and its unequal health impacts (for instance certain nationalities or social classes) to more local matters such as unequal access to green spaces. The environmental justice movement has certainly developed further in the USA than most other countries, especially compared with the UK, but what it captures or embodies is a mixture of environmentalism, environmental ethics or philosophy, together with concern for local community health and global public health. In the main, however, environmental justice is less about the health of the environment per se (its intrinsic worth) but the health of the environment in relation to human use and its impact on human health – its instrumental value.

It is now more apparent how our last conception of air and health fits in, and what it means. Part IV started by looking at climate change, its scientific basis and the development of interest in its causes and consequences. It is now widely accepted that climate change has arisen due to man-made pollution accompanying industrialisation and modern Western development. The sequelae of climate change will impact on the health of the planet, and of humans, now but especially in the future. The causes and consequences of climate change are unequally, and unfairly, distributed. Calls for 'climate justice' have been increasing in reaction to these inequities, as well as due to the slowness of progress in policy around mitigation and adaptation.

What this last conception of air and health truly embodies, however, is an accumulation of discontent around Western living, Western values, care of the environment and global poverty – along with displeasure with policy-makers' efforts to address these. Climate change captures the dissatisfaction of environmentalists and environmental ethicists with the way the planet and its natural resources (which include air) are treated, and is often used as an example or case study in books and journals of these disciplines.

Climate change, however, still pushes forward anthropocentric views of nature, through its emphasis on the impact on humans. In this sense it is strange to find it in the environmental activist's toolkit but it is there because it is a powerful example of what happens when nature is inappropriately valued. It is there because it can further an important agenda and because it represents growing disillusionment with the way things are.

But changing the way things are has proven hard, as stumbling climate change policy has shown, because the roots of our current problems lie very deep, and have become ingrained in Western lifestyles. To have purchase, attempts to tackle seriously problems such as climate change need simultaneously to address the roots that have bred liberal individualism, dominant utilitarian-based political philosophy, materialism and social atomism. Much of this is echoed in the disdain that modern virtue ethicists, and others such as Wittgenstein, have shown for the direction of contemporary Western moral philosophy.

For environmental work within public health, similar arguments apply. As a subsection of public health, environmental health has always been concerned with the impact that the environment has on human health. However, not only is the environment here considered as an instrument, a means, to human well-being, but environmental health largely encompasses man-made damage to the environment and the subsequent effects on humans: contamination of the land with hazardous chemicals; landfill sites; factory products; unwanted effluents into water supplies; and, of course, air pollution as addressed in Part III.

More recently, through a combination of attention to inequalities in health and general environmentalism, we have seen growing attention to so-called environmental injustices, whether these be global matters, such as climate change, or more local concerns such as residents in deprived areas living disproportionately close to industrial pollutants. These are important and valuable advances, but in Naess's language they reflect shallow environmentalism, touching only the surface of the problem, and unlikely to yield substantial results. Deep environmentalism, in contrast, requires commitment to radical cures such as significant personal and cultural change.

So far the climate-change debate would support Naess's views. Without addressing the deep roots of the current environmental crisis, without valuing the environment for its inherent worth, real progress in protecting the environment – and indirectly human health – may be hard to achieve.

NOTES

1. In parallel with developments such as Francis Bacon's (1561–1627) inductive science.
2. The rulers, or guardians, are philosophers who reach that position on their own merit but whose reward needs to be satisfaction of doing their job well, rather than monetary.
3. Psychological egoism presents that people do, in fact, act in their own self-interests but is different to philosophical egoism which argues that people *should* behave in this way. The two are often confused, and the former used as justification for the latter.
4. The difference between having interests and goals has been stressed by philosophers. A plant may grow toward light, or a tree may wither and die, but neither the plant nor the tree, arguably, has attitudes towards these happenings.
5. The distinction between living and non-living is often neither biologically or philosophically clear. For instance, a rock may be considered non-living or inert but what about soil?
6. These are non-maleficience (to any organism), non-interference, fidelity (to not betray or deceive wild animals) and restitutive justice (to restore balance if harm is done).

REFERENCES

1. R. Descartes, *Discourse on Method and The Meditations* (trans. F. E. Sutcliffe) (London: Penguin, 1968). [Originally published separately in 1638 and 1641.]
2. R. S. Westfall, *The Construction of Modern Science: Mechanisms and mechanics* (Cambridge: Cambridge University Press, 1977).
3. T. L. Hankins, *Science and the Enlightenment* (Cambridge: Cambridge University Press, 1988).

4. M. Midgley, *Science and Poetry* (London: Routledge, 2001).

5. W. Kymlicka, *Contemporary Political Philosophy* (Oxford: Oxford University Press, 2002).

6. W. K. Guthrie, *The Greek Philosophers: From Thales to Aristotle* (London: Routledge, 1989).

7. M. Cohen, *Political Philosophy: From Plato to Mao* (London: Pluto, 2001).

8. J. Rachels, *The Elements of Moral Philosophy* (New York: McGraw-Hill, 1993).

9. J. J. Rousseau, *The Social Contract and Discourses* (trans. G. D. H. Cole) (New York: Dutton, 1959). [*Of the Social Contract* originally published in 1762; *Discourses* originally published separately in 1754 and 1755.]

10. M. Midgley, *Science and Poetry* (London: Routledge, 2001), p. 159.

11. J. Passmore, *Man's Responsibility for Nature* (London: Duckworth, 1974).

12. R. Dawkins, *The Selfish Gene* (Oxford: Oxford University Press, 1976).

13. M. Midgley, *Science and Poetry* (London: Routledge, 2001), p. 69.

14. A. Light and R. Holmes, III, eds., *Environmental Ethics: An anthology* (Oxford: Blackwell, 2003).

15. R. Elliot, Environmental ethics. In P. Singer, ed., *A Companion to Ethics* (Oxford: Blackwell, 1997), pp. 284–93.

16. K. Goodpaster, On being morally considerable. *J. Phil.*, **75** (1978), 308–25.

17. J. B. Callicott, Elements of an environmental ethic: moral considerability and the biotic community. *Env. Ethics*, **1** (1979), 71–81.

18. J. R. Des Jardins, *Environmental Ethics: An introduction to environmental philosophy* (Belmont: Wadsworth, 1997).

19. P. Singer, *Practical Ethics* (Cambridge: Cambridge University Press, 1999), p. 285.

20. —, *How Are We To Live? Ethics in an age of self-interest* (Oxford: Oxford University Press, 1997).

Conclusions to Part IV

This part of the book has looked at a last conception of air and health, the approach to dealing with climate change, and has explored what the conception represents and what the conception tells us about current and future prospects for public health. Chapter 11 has also tried to bring together ideas discussed in the previous chapters, and the different conceptions of air and health within those.

Climate change is the package of climatic consequences of greenhouse warming, a heating up of the atmosphere resulting, in the main, from 150 years of industrialisation. The countries most responsible historically for the causes of greenhouse warming are Western but over the past few decades the contribution from industrialising low- and middle-income countries has become significant. While climate change is a global problem, it is also a global public health problem as the health consequences are severe, and are likely to impact differentially on poorer countries with their limited abilities to cope or adapt. Policy developments around climate change have stuttered, with little substantial advance to date.

Within the climate-change debate there has, however, been a distinct focus – academically and in policy – on the unfair, or inequitable, distribution, of the causes and consequences of greenhouse warming. This focus on equity has also involved an activist movement, which one could collectively call climate justice, involving a combination of intellectual work, debate, websites and advocacy. The climate justice movement reflects a general growing interest in health inequalities, and health inequities, which have drawn to significant degree on John Rawls's ideas about justice and fairness. In public health in the UK there is, for the first time, a definitive commitment to redressing the problem of inequitable distribution of health and inequitable access to health care, largely through attention to improving the lot of those worst off.

Not only, however, does climate change reflect interest in health inequities but it also embraces contemporary concerns about the plight of the environment. The activist component of climate justice is an extension of the environmentalism of the 1960s and 1970s, which has been accompanied by the emergence of the field of environmental ethics. However, while commitment to reducing inequities is laudable, as is commitment to improving the environment, environmental ethics reminds us that the roots of current environmental ideas lie deep. Several hundred years of separation in Western thought of mind from matter, subject from object, and values from facts, has resulted in the dominance of scientific reductionism over holism, and the devaluing of nature. Connected developments in moral and political philosophy have ingrained utilitarianism and liberal individualism, justifying self-interested behaviour and leading to social atomism. The depth of the problem means meaningful solutions need to be radical.

For public health, the environment has predominantly been of instrumental interest, as it relates to concern with the human health consequences of environmental damage. While environmental matters such as chemical hazards and even outdoor air pollution are undeniably important, they really only attend to the superficial end of the spectrum, representing shallow environmentalism. For the future of the planet and its inhabitants a more substantial change in attitudes is required.

12

Conclusions and recommendations

This book has examined the theme of air and health over a period of more than 2000 years. Two main aims were set out in the Introduction. The first was to explore the changing conceptions of air and health over two millennia, alongside – in more recent times – developments in public health in the UK. The second, related aim, was to look critically at how such changing conceptions might illuminate contemporary issues in public health theory and practice; and, also, what the changing conceptions of air and health might reveal about today's environmental problems.

To undertake the challenge, a range of academic disciplines has been drawn upon: the history of medicine and health policy; moral and political philosophy; science; environmentalism; and environmental ethics. Different approaches have also been used. At times it has been necessary to look at broad links, for instance between ancient civilisations; on other occasions it has been important to concentrate on the historical detail of a certain period, for instance public health in England in the second half of the nineteenth century.

The difficulties posed by definitions have needed careful negotiation. As has been discussed through the course of the book, public health can be broadly treated as encompassing collective actions aimed at improving the health of the public. But even then, as Hamlin points out, questions of what we mean by 'health' and 'public', and what is understood as the proper domain of public health, are all contested matters. Broken down, public health may be considered as profession, science, component of public administration, and vision [1]. As another example, the term 'environmental health', as used by public health professionals, has tended to relate to how elements of the natural environment affect human health; environmental health is interpreted rather differently by environmentalists.

Conclusions

The main findings of this book, and what conclusions can be drawn from them, can initially be brought out by reviewing the book. Part I charted changes in the conception of air and health from ancient civilisations to the bacteriological era at the end of the nineteenth century. In Egyptian medical thought, air had a special supernatural place as the creator and sustainer of life, not dissimilar to the life-force *qi* in Chinese medicine, or early Judeo-Christian vitalism, in which the breath of life lay in air, imbued with the Spirit of God.

Greek medicine attempted to remove religious elements and provide the first rational medical theory, but supernatural ideas remained, as the concept of *pneuma* connected air with the soul in the *Hippocratic Corpus*, and also later in the works of Plato and Aristotle. As well as emphasising empirical observation, Greek medicine was naturalistic, understanding mankind as part of nature, and illness as a natural phenomenon. Balance within the body, and between the body and the natural environment maintained health, and imbalance resulted in disease. Air was part of the natural environment, and epidemics were felt to be carried by polluted air or miasma. But air was also internalised, connected in this capacity with specific illnesses, such as the sacred disease, now known as epilepsy.

The conception of air and health in Greek medicine embodied a holistic relationship both within the human body, and between humans and the natural environment. This conception continued with Roman medicine, and through to the Enlightenment. By that time, however, air was at the centre of a debate about the causation of infectious diseases. A spectrum of beliefs about disease causation existed, incorporating at one end the idea that poor sanitary conditions created polluted air – miasma – that was responsible for epidemic outbreaks; and, at the other end, minute particles, contagion, were held to be the cause – or a combination of the two explanations was invoked. The terms miasma and contagion were used unpredictably, sometimes to answer different questions, but air was the medium of disease in both.

This debate about disease causation was framed by a wider debate through the nineteenth century, about the role of the environment in shaping biological and social evolution. Professional public health emerged in the middle of that century with an emphasis on improving the unsanitary environmental conditions of the poor. The context was the need for a healthy workforce at a time of economic expansion, but to some the immoral poor were responsible for their adverse living conditions, and assistance would run counter to moral progress.

As scientific rationalism began to underpin medicine, the relationship between air and health lost the spiritualism and holism of earlier epochs. Though atmospheric pollution had troubled early civilisations, from around the seventeenth century onwards air became gradually equated with polluted air, first as miasma then more visibly as smoke pollution. As covered in Part II of the book, naturalism also disappeared from the conception that was developing, of man-made polluted air and its effects on human health. Industrialisation in the cities led to increasing output of smoke from factories and domestic coal fires. The atmosphere darkened, and air in relation to health corresponded with the adverse effects of the smoke-filled skies on human well-being: deaths; respiratory illnesses such as pneumonia and TB; and psychological effects of the gloom.

As air became reduced to the polluting component of smoke, the statistical search for associations between smoke pollution and mortality or morbidity began. This was repeatedly demonstrated through the early twentieth century but, despite the evidence, and in the face of a growing anti-smoke campaign, policy to reduce smoke levels failed to deliver. The developing public health profession, and MOHs in particular, worked with campaigners and communities, advocating improvements to environmental and atmospheric conditions.

The severe smog in London in 1952 was possibly the straw that broke the camel's back, in terms of leading to change in policy. The severity of that episode, along with public concern and mounting pressure from campaigners, led to a significant legislative event, the Clean Air Act. But this act was a watered-down version of the Beaver Committee's recommendations, in particular the act's failure to cover domestic smoke pollution. The fall in smoke production that was to ensue over the following decades probably had more to do with falling prices of alternative fuels, as well as a growing emphasis through the 1960s on personal responsibility for matters relating to health.

The period between the world wars has been referred to as the heyday for the (medical) public health profession in Britain, with authority and control at their highest. The profession, however, lost influence over significant medical services with the formation of the NHS in 1948, and its powers were further diminished by the removal of control over sanitary officers and social workers. As public health moved into health authorities, connections with local authorities (where environmental health officers would be located) were attenuated, as was the ability of the profession to engage substantially in matters of environmental health.

The status of public health in the UK may have been bolstered in the 1970s through the formation of the Faculty of Community Medicine, and through recognition as a medical speciality, but the costs included increased managerial responsibilities over provision of health services, and marginalisation of those not medically qualified. For public health, improving population health became dominated by improving health services, with less emphasis on social determinants and the environment.

During the second half of the twentieth century, the conception of air and health as polluted air and its effects on human health continued. But polluted air was gradually reduced to its constituent components. First smoke and sulphur dioxide, then oxides of nitrogen, ozone, and particulate matter. Particulates have been further divided by diameter, with differential health impacts associated with different diameter particles. As considered in Part III, the epidemiological search for associations between components of polluted air, and mortality or morbidity, has been extensive over the past two decades.

The example of quantitative risk assessment (QRA) illustrates that this reductionistic orientation has been driven by technological and methodological advances – in the measurement of pollutants and health effects, and in data handling and statistical analysis. But the QRA case study also brings out some of the constraints of modern epidemiology: the limitations of 'black-box' thinking and the focus on proximate risk factors; the dominance of evidence-based medicine and evidence-based policy, and the limitations of their positivist nature; the lowly status of cross-sectional studies within hierarchies of what counts as evidence; and the lack of theoretical development around population health.

The approach taken in dealing with climate change represents the final conception of air and health, and was explored in Part IV. Industrialisation and Western lifestyles have led to a warming of the atmosphere, which has resulted in meteorological changes including a rise in average temperature, more extremes of temperatures, and increased incidence of events such as heatwaves and floods. These climatic changes will have a variety of adverse health effects, direct and indirect, although there may be some health benefits. The negative health effects, and the ability to adapt or mitigate against them, are likely to be unequally borne by rich and poor communities and countries.

This last conception raises interesting questions about the philosophical foundations of public health, and the links between these foundations and other developments in the history of science and moral philosophy. Utilitarianism has traditionally been the foundation of public health practice. But climate change illustrates the limitations of utilitarian thinking: the focus on proximate impacts (many health effects will occur at a distance from the emitters of greenhouse gases); the presentism (many health effects will occur well into the future); the reliance on empirical data; and the difficulties of taking sufficient account of equity (fair sharing of the burdens of climate change).

Social justice, as espoused by its leading proponent John Rawls, has emerged as an alternative to utilitarianism, arguing for the more equitable distribution of societal burdens and benefits. It remains, however, a contractarian approach, and has been thrust into the climate-change debate to press for fairer mechanisms of dealing with the costs (and other burdens) of the damage from greenhouse warming. Extensive academic work around equity issues has informed the climate-change policy debate, buttressed by the climate justice movement with its environmentalist and activist elements.

Social justice has also, quite recently, fed into national policy in the UK around reducing inequalities in health, though in practice such reduction may be hard to achieve in the context of economic policies that drive inequalities apart. Developments that take more account of equity may indeed be laudable and progressive but there are other dimensions to seeing social justice as the main alternative to utilitarianism.

For instance, the direction that Western moral philosophy has taken in the twentieth century (and thereby the moral theories that have been developed) is, to some, questionable. One of the most renowned modern philosophers, Ludwig Wittgenstein, deliberately steered clear of moral philosophy, despite his works touching many other aspects of philosophy. Wittgenstein felt that ethics was a deeply personal matter, about issues such as integrity, honesty and how one lives one's life. He had little time for so-called experts, professional moral philosophers, and was troubled by the abstract focus on theoretical development within modern Western moral philosophy [2]. In a similar vein, proponents of virtue ethics, such as Alisdair MacIntyre, have criticised how Western moral philosophy has centred on utilitarian and deontological theories, as indeed has Western bioethics. There has been a call for rethinking the direction of moral philosophy, with greater attention to development of Aristotelian ideas around the virtues, excellences of character, and their place in the space between individual fulfilment and social good [3].

As is argued in Part IV, historical developments in moral philosophy are, however, linked to developments in political philosophy and the history of science, and understanding how we value the environment today needs bringing these together. The Cartesian separation of mind from matter, value from fact, has impacted in connected ways. In scientific medicine the body, stripped from the environment and earlier holistic conceptions, has become a material entity within which disease may occur. Progressive reductionism has demarcated the body into separate sections, corresponding to different specialities.[1]

In political philosophy, the concern of the Greeks for the well-being of the whole (the city-state), and how this meshed with individual flourishing, has been largely lost – within Western capitalism at least. The brutal world of Thomas Hobbes oriented morality around a social contract needed for self-protection and, together with Locke's attention to ownership rights, heralded the individualism that is now the corner-stone of modern liberalism.

Importantly, any way of seeing the world entails a belief about that which should be valued, a moral component. The way we think about our material selves is connected to how we think about our duties to other humans, animals and the natural world. The accepted emphasis on the individual, and his or her rights, has shifted duties towards others to the responsibility of the State. At the same time, manipulation of the natural environment has become understood as fundamental to individual and utilitarian aims. But reducing nature to an instrument for human needs has dramatically changed the way we see the natural world.

Environmental ethicists would argue that this slow shift, from perceiving the natural environment as having inherent value to its having instrumental value (to human needs), is at the core of the current environmental crisis. But the roots of the problem lie deep, and resolution will require considerable effort, and radical reform. The climate-change debate illustrates how most current efforts at environmental improvement really just scratch the surface.

Mary Midgley provides the image of a sinking terrestrial ship of environmental problems [4], which is not, at present, sinking at the end bearing Western, high-income countries:

Of course it is understandable that we do not see the planetary danger. Other, more immediate evils constantly demand our attention. Conditions on the terrestrial ship are bad in a thousand ways and endless things need to be done about them. But if the ship sinks, curing those evils will not be much help. The message is not that we should value the health of the earth above human needs. It is that these are not alternatives. Without a healthy earth humans cannot survive anyway.

Implications and recommendations

This book has concentrated on Western cultures, in particular Western medical ideas and public health in the UK. The recommendations naturally have special importance to the UK but theoretical aspects, and their practical implications, are of general relevance. Before discussing these recommendations, it is therefore worth outlining developments in public health in some countries outside the UK. Current developments in UK public health were discussed in the Introduction.

Public health in countries outside the UK

Notwithstanding the difficulties around definition discussed earlier, public health has progressed in different ways in different countries, only sometimes involving the distinctive form of professional public health that has evolved in the UK. Countries selected represent a variety of different histories.

Like the UK, Sweden has had a strong tradition in public health. Vital statistics and death registration were initiated in Sweden in 1749, the earliest for any European country, and provincial medical officers were required to submit annual reports on the health of their populations from 1854. It is, however, the renewed interest in public health in Sweden since the 1970s that is especially salient. Sweden's first national *Public Health Report* was published in 1987, following a recommendation by the Health Care Commission – accepted by the Government – that an analysis of the health of the population should be produced every three years. In 1997 the Swedish government appointed a National Committee for Public Health with responsibility for creating public health strategies and goals, and the new millennium has seen continued commitment to the public health vision, with an emphasis on good health for all on equal terms. Investment in public health education, research and training is reflected in the 18 proposed goals for the Swedish National Public Health Policy [5].

By contrast, public health in eastern Europe and the former Soviet Union has had a difficult time over the past few decades. In the second half of the twentieth century population health has, in general, either worsened or stagnated in most countries; summary statistics, such as life expectancy at birth, have sometimes concealed deterioration in, for example, mortality among young men, through the counteracting improvement in infant mortality. McKee and Watonski describe how a career in public health has generally been an unattractive option in the former Eastern Bloc, and public health capacity has been poor in a range of settings – public health organisations, government, academia, and in non-governmental organisations. In the Soviet Union particular problems in public health development arose due to the intransigence of the sanitary–epidemiological model, and as a result of Marxist–Leninist teaching that has labelled emerging health threats as temporary, and attributable to the transition to communism. However, the outlook for public health capacity now appears less bleak: new schools of public health have been created in Hungary and Croatia, while networks of academic centres include the Baltic republics and parts of Russia, supported by developing links with countries such as Israel and the UK [6].

In North America the histories of public health, including its professional development, differ significantly between the USA and Canada. In the USA, public health has been, until relatively recently, mainly a local matter. As Fee explains, this can be attributed to the economic, political and cultural complexities of the formation of a new nation, combined with the development of different geographical regions of the new nation at different times [7]. Early US public health was through town-appointed inspectors and was largely concerned with control of infectious diseases. During the industrialisation of the late nineteenth and early twentieth centuries public health was more urban, and focused on infectious diseases. After the Civil War most states created boards of health, and in 1872 the American Public Health Association was formed. The Federal Government created the US Public Health Service in 1912, initially worried about the disease status of the immigrant population.

Though new schools of public health – such as Johns Hopkins – were set up in the early decades of the last century, non-medical students predominated, as clinicians were sought by state health departments regardless of their public health qualifications. Fee describes that this structural problem in the relationship between medicine and public health has endured in the USA: public health today may be multi-professional but a special status is accorded to those who are medically qualified. In some ways this is reflected in the new speciality of preventive medicine which appeared after the Second World War, directed towards bridging the gap between public health and the clinical sciences, especially around non-communicable diseases, but also further privileging doctors [7].

Despite the important 1988 US Institute of Medicine Report *The Future of Public Health* [8], in which a list of ten essential public health functions was described, local and state level public health remains variable around the USA, and heavily involved in the provision of medical services. With limited investment in public health infrastructure over recent decades, and no formal structure for education and training, there has been an erosion of public health departments, with high staff attrition rates. Public health schools are more involved in training those working in the federal sphere or with international agencies.

In Canada it could be argued that, despite an uneven history, public health has, in recent times at least, been more prominent [9]. The overall health experience in Canada has generally been more favourable than in the USA, as measured by indicators such as life expectancy and infant mortality, and inequalities in both income and health are less dramatic in Canada than the USA. Due to Canada's comprehensive national insurance programme, public health activities have not been dominated by issues of equity of health service provision. As a result, modern public health services – local, provincial and federal – have been able to focus more on the underlying determinants of health. Canadian public health has spearheaded initiatives such as 'healthy cities, healthy communities' and the Ottawa Charter for Health Promotion [10], and the establishment of the National Forum on Health in 1993. However, as in other countries, concerns remain about the strength of the public health infrastructure in Canada, although the terrorist events of 11 September 2001 may have reinforced the importance of public health [11].

Public health in Australia and New Zealand, as in Canada, has been heavily influenced by the British tradition [12]. By the beginning of the twentieth century, public health legislation had been passed in both countries establishing (local) government agencies with responsibilities around environmental health and regulation. As discussed by Davis and Lin the role of the MOH (around infectious disease control, food safety and hygiene, and aspects of living and environmental conditions) was central to early public health activities, and remains important today [13]. In the second half of the last century central government responsibilities have grown, for instance around health service provision, and in New Zealand the poor health of the Maori people has been a key, culturally sensitive concern. In both countries public health funding in recent decades has remained in line with other developed countries, although the creation of the Public Health Commission in New Zealand in the early 1990s temporarily suggested more substantial commitment. Public health workforce development has been relatively strong, especially around post-graduate programmes (educational and practice-based) in public health in the 1980s and 1990s, open to clinical and non-clinical graduates, and promoting a progressive approach to multidisciplinary public health.

With an outline of developments in public health in different countries in mind, we can now return to the findings of the book, the recommendation of which are applicable to public health education, training and practice in different settings.

Public health theory and philosophy

There remains no theory of public health on which the practical doing of public health can be based. There are two reasons why this may be so. First, it could be that public health needs no theory of its own or, at least, can function reasonably without such a theory. Alternatively, it could be that insufficient attention has been given to drawing together a cohesive theory. These will be looked at in turn.

Almost all medical or health systems have conceptual frameworks or theoretical foundations on which they are based. Western medicine is founded on a biomedical model of health, more recently extended to some degree to a bio-psycho-social model. The basic biomedical sciences of anatomy, physiology, biochemistry and pharmacology underpin an explication of health and disease. The depersonalised individual's body is the focus. Similarly, almost all the many so-called complementary therapies – homeopathy, acupuncture, Ayurvedic medicine, herbalism and so on – have their own theoretical foundations. These may be far removed from those of scientific medicine but they are theoretical foundations nonetheless.

All these systems are about the health of the individual, albeit with some degree of attention to broader aspects that may explain individual health. Public health is different because it is about the health of communities. But communities are healthy or ill for various reasons, so it would follow that public health merits a theoretical foundation as a basis for understanding and action, just like all these other health models do. Yet none really exists, leaving practitioners to draw eclectically, and somewhat in the dark, on a mixture of theoretical ideas and foundations [14].

To counter this it could be argued that, unlike models of individual health, public health is about political will, cultural ethos and collective action, making a theoretical base somewhat redundant. But this misses an important point. Models of individual health are culturally bound and determined no less than any putative model of population health. Western medicine is inextricably linked to the values of the scientific enterprise, and to the needs and interests of those served by taking society, nature, the environment, God and even the person out of the explanatory jigsaw puzzle of health, and making the human body the largely passive recipient of misfortune and chance. If public health is clearly set in its cultural context, so too are models of individual health. It is just that the connection between sociopolitical values and public health is more obvious.

So what might it mean to have or develop a theoretical foundation for public health? This naturally depends in part on what one means by public health, but at an academic level consists, at a minimum, in theoretical development around explanation(s) of what makes communities or populations ill or healthy. Any such theory needs to incorporate differences between communities or populations, and must embrace determinants most upstream (or distal) to communities, as well as those closest, most downstream (or proximal).

To date there has been a paucity of theoretical work around population health. The most recent example of an attempt is probably Krieger's work on an ecosocial framework for public health described in Part III. But the field is otherwise fallow, leaving public health to deal with matters of intuitive importance or political determination, rather than those arising from critical development of ideas and imagination.

Recently Weed has called for much more work to create a philosophy of public health, combining a true epistemology of public health with an ethical framework [15]. The first recommendation of this section echoes Weed's sentiments but does not suggest that this will be an easy task. It will require academic commitment, ignited possibly by an international workshop or conference on philosophy of public health. As a start this would highlight existing activities and intellectual ideas, from which coordination of further work could begin.

Ethics and public health

The findings of this book suggest that there is significant scope for further work around ethical issues in public health. In fact, over the past decade or so, a new term, or field of enquiry, has emerged called 'public health ethics'. Although first coined in the USA, public health ethics has begun to appear in the UK public health professional literature. However, what public health ethics means, and how it should be approached, needs careful consideration, as I described in an article that was published in the journal *Social Science and Medicine* in 2003 [16].

In that paper I argued that ethics today certainly appears to be in vogue [17], and the last four decades have witnessed a corresponding increase in interest in ethics in medicine and the natural sciences. To avoid being another step along what the medical historian Cooter has called the 'resistible rise of medical ethics' [18], public health ethics needs to meaningfully contribute to debate in the field.

Following a literature review of articles on public health ethics, accompanied by a questionnaire survey of public health ethics teaching in the UK, I concluded that public health ethics can be interpreted or approached in three broad ways. The first, most common, concept of public health ethics is as a list of topics or problems in public health, all of which have a moral dimension [19]. Most of these analyses, however, tend to be superficial and may tell us more about the types of questions preferred in the mainstream, and those sidelined or avoided.

Another way of thinking about public health ethics is as a branch of biomedical ethics that draws on the analytical tools of that discipline. As examples, the *Oxford Textbook of Public Health* [20] and the *Public Health Primer* [21], as well as two key American textbooks [22, 23], discuss public health ethics in terms of analysis of problems in public health using principles and frameworks developed in biomedical ethics. This kind of interpretation does have benefits. The use of a set of principles can be helpful in providing a platform from which to start, and can serve to concentrate the mind and focus attention. Shortfalls, however, with this approach begin with the position and value one puts on contemporary biomedical ethics in the first place. While some consider medical ethics as moral philosophy applied to medicine, other less familiar approaches exist such as care ethics [24], feminist and social science perspectives on medical ethics [25], casuistry and historical critiques [26]. Yet public health ethics tends to draw on one particular approach only – the four principles[2] of biomedical ethics [27, 28], and this approach has well-rehearsed limitations[3] [29]. When public health ethics uses principlism as its analytical framework, the same criticisms can be levelled.

One leading figure in contemporary medical ethics is against principlism for two reasons that are highly relevant to public health: its individualistic bias (the dominance of the principle of respect for autonomy) and its blocking function. As a self-described 'communitarian philosopher', Callahan is interested in the social meaning of ethical problems and the wide impacts of policies on societies, and is particularly concerned by the impeding function of principlism [30]. He describes the restrictive impact of principlism [31]:

Instead of inviting us to think as richly and imaginatively about ethics as possible, in fact it [principlism] is a kind of ethical reductionism, in effect allowing us to escape from the complexity of life, and to cut through the ambiguities and uncertainties that mark most serious ethical problems.

Public health is about communities rather than individuals, is full of complexities, and Callahan's concerns are directly relevant. To avoid association with the principle-based approach of biomedical ethics, use of the term 'public health ethics' might best be avoided. A more helpful approach would be to think about philosophy of public health or the exploration of ethics in public health.

This book supports the view that values permeate all areas and levels of public health theory and practice, and are culturally bound. Moral and philosophical analyses of problems are key, but historical, sociological and anthropological approaches to the examination of ethics in public health are equally relevant. Different methodological approaches would then allow for a truly interdisciplinary approach to exploring ethical issues in public health, and the importance of environmental ethics to public health could be recognised.

The exploration of ethics in public health is at a fairly early stage. It is essential that critical debate and research continues, connected to development of a philosophy of public health. Some programmes have recently been established that concentrate on research around ethics in public health,[4] and progress will also be fostered through education in public health, as is discussed below.

Public health education and training

Those working in public health in the UK and many other countries come from a variety of professional and academic backgrounds including medicine, epidemiology, statistics, social science, health promotion, and the allied healthcare professions such as nursing and psychology. Many of these will have undertaken postgraduate education – usually as a Masters degree or Diploma in Public Health, or equivalent – prior to, or in parallel with, their involvement in public health work.

Formal training programmes in public health are being increasingly advocated in the UK for all public health practitioners and, as covered earlier, are being similarly developed in other Western countries. Traditionally the domain of doctors working in public health, these training programmes are being opened up to non-medical practitioners, or equivalent programmes are being developed. A minimum educational component of such programmes tends to be participation in a Masters degree in Public Health (MSc or MPH),[5] now offered by many institutions in different countries. These degree courses naturally vary but in the UK their content generally overlaps with the syllabus for the first part of the examination for Membership of the Faculty of Public Health (MFPH).

For those planning to work in public health in the UK, the MFPH examination is becoming an increasingly necessary part of career advancement. Originally aligned with other medical speciality examinations, it currently consists of two parts: MFPH Part I and MFPH Part II. The Part I examination tests mainly theoretical aspects of public health and is taken relatively early in training. The Part II examination addresses competency at practical aspects of public health work and is taken further along the training pathway. The examination was opened up to non-medical public health specialists in 2000.

The Part I examination tests knowledge of the syllabus (Box 12.1), and also tests skills at using and interpreting this knowledge as well as skills relating to communication and presentation (Box 12.2). As the summary headings in the boxes show,

Box 12.1 Diploma and Part I MFPH exam syllabus[6]

a. Research methods (epidemiology, statistics, needs assessment, health economics, qualitative methods)
b. Disease causation and prevention, and health promotion (specific diseases and their epidemiology, screening, genetics, health and social behaviour, environment, communicable diseases, principles and practice of health promotion)
c. Health information
d. Medical sociology, social policy and health economics (includes some aspects of equality, equity and social justice)
e. Organisation and management of health care.

Box 12.2 Skills tested at Diploma and Part I MFPH exam

a. Design and interpretation of (research) studies (includes quantitative and qualitative research)
b. Data manipulation and interpretation
c. Communication: written presentations skills; preparation of papers for publication; preparation of material for different audiences. . . ; information handling; use of media . . . disease prevention and health promotion.

the breadth of theoretical knowledge felt to reflect what is needed to work in public health appears wide-ranging. Nevertheless, the emphasis of the syllabus remains centred on the scientific aspects of public health (epidemiology, statistics, health information), medical aspects (communicable diseases), and the organisation and management of health services. As important as these undoubtedly are, this book has argued that there are deep-seated reasons for the changing Western understanding of health and the environment, with implications for our comprehension of the health of populations and the development of public health practice. Within public health there appears to be a lack of awareness of the historical, philosophical and political roots of public health; and insufficient attention to the environmental and social determinants of health, despite the basics of environmental health being covered in the syllabus.

So a gap may exist between the knowledge needed, and that currently taught, a gap that probably holds for public health postgraduate courses or examinations in many countries. The main areas that are missing or deficient in public health education appear to be:

- medical humanities (including ethics) and public health;
- history and philosophy of science, and public health;
- history of public health;
- political science/political philosophy, and public health;
- environmental health: environmental determinants of human health, and also the health of the environment itself.

Identification of gaps is, however, the easier part of changing the nature or content of public health education. Even if there were agreement that these are important areas that need covering, difficulties exist such as lack of space in the curriculum. Space, however, could be made by reducing the amount of teaching oriented around the scientific dimensions of public health, and also reducing elements concerned with the organisation, management and provision of healthcare services. This latter aspect is particularly symbolic because, over the last few decades, improving health has become increasingly equated with improving healthcare services. As was discussed earlier, professional public health has correspondingly enhanced its involvement in the provision of services, at the expense of tackling the root causes of what makes communities unhealthy. It is an opportune time to redress the balance.

Another difficulty tends to be shortage of skilled staff to do the teaching. Educators in public health tend to come predominantly from scientific or medical backgrounds, making it hard to provide education in areas such as the medical humanities, ethics in public health, and political philosophy. Finding staff to teach the new areas may prove hard at the outset but should become easier with time.

The gap between academia and practice in public health

Bridging the gap between academic and practical work in public health has proved difficult. For practitioners, short- and medium-term priorities tend to dominate, often shaped by political priorities. When an academic input is sought, it is likely to be about research with immediate practical significance, often around an issue of health service provision. However, the current emphasis on areas such as health inequalities and health impact assessment may herald new ways of working together, which could be supported by education and training that fostered deeper theoretical understanding.

An onus exists too on those working in academic public health to become more involved in what is important at the ground level. Accusations of 'ivory tower' academia are often unfairly cast, and it is true that pressures within universities on funding, on doing research in prescribed areas and publishing findings in preferred journals, all place limits on what academic public health, in the UK at least, can realistically achieve [32–34]. Nevertheless, public health academics do need to be willing to apply their skills to areas of practical relevance. This does not run counter to the recommendations that this book has made about the need for more work around ethics and philosophy in public health, but stresses the importance of theoretical development in public health, alongside academic involvement in areas of practical importance.

Public health advocacy

A final word needs to be said about advocacy within public health, a vital role that seems to have diminished over time. Whereas the MOH provided a voice for the local community, and proponents of social medicine challenged political and economic considerations, the modern public health professional has become less of an advocate for change.

While the demands placed on those working in public health today are undoubtedly high, the time is ripe for public health academics and practitioners, together, to re-invigorate the role as campaigner and advocate for social reform.

NOTES

1. Parallel reductionistic tendencies have occurred in epidemiology, as discussed in Part III.
2. Developed by the American philosophers Beauchamp and Childress in the late 1970s, 'principlism' uses four principles (respect for autonomy, beneficence, non-maleficence and justice) as a basis to address moral dilemmas in medicine.
3. These limitations include lack of rational explanation of why these principles have been chosen, not having a means of weighing up competing principles, Western bias and failing to give sufficient attention to the context of dilemmas addressed.
4. For example: Program in the History and Ethics of Public Health, Department of Sociomedical Sciences, Joseph P. Mailman School of Public Health, New York, USA; International Programme in Ethics, Public Health and Human Rights (IPEPH), London School of Hygiene and Tropical Medicine, UK.
5. Or similar degree.
6. This box has been drawn from the website of the Faculty of Public Health Medicine (UK), www.fphm.org.uk (accessed 25 June 2004).

REFERENCES

1. C. Hamlin, The history and development of public health in developed countries. In R. Detels, J. McEwen, R. Beaglehole and H. Tanaka, eds., *Oxford Textbook of Public Health*, 4th edn (Oxford: Oxford University Press, 2004), pp. 21–37.
2. C. Elliott, ed., *Slow Cures and Bad Philosophers: Essays on Wittgenstein, medicine, and bioethics* (London: Duke University Press, 2001).

3. R. Crisp, ed., *How Should One Live? Essays on the virtues* (Oxford: Oxford University Press, 1999).

4. M. Midgley, *Science and Poetry* (London: Routledge, 2001), p. 206.

5. S. Wall, G. Persson and L. Weinhall, Public health in Sweden: facts, visions and lessons. In R. Beaglehole, ed., *Global Public Health: A new era* (Oxford: Oxford University Press, 2003), pp. 70–86.

6. M. McKee and W. Zatonski, Patterns of health in eastern Europe and the former Soviet Union. In R. Beaglehole, ed., *Global Public Health: A new era* (Oxford: Oxford University Press, 2003), pp. 87–104.

7. E. Fee, Public health and the state: the United States. In D. Porter, ed., *The History of Public Health and the Modern State* (Amsterdam: Clio Medica, 1994), pp. 224–75.

8. Institute of Medicine Committee for the Study of the Future of Public Health, *The Future of Public Health* (Washington, DC: National Academy Press, 1988).

9. J. Cassel, Public health in Canada. In D. Porter, ed., *The History of Public Health and the Modern State* (Amsterdam: Clio Medica, 1994), pp. 276–312.

10. World Health Organization, A charter for health promotion (the Ottawa Charter). *Can. J. Pub. Health*, **77** (1986), 425–30.

11. F. D. Scutchfield and J. M. Last, Public health in North America. In R. Beaglehole, ed., *Global Public Health: a new era* (Oxford: Oxford University Press, 2003), pp. 105–19.

12. L. Bryder, A new world? Two hundred years of public health in Australia and New Zealand. In D. Porter, ed., *The History of Public Health and the Modern State* (Amsterdam: Clio Medica, 1994), pp. 313–34.

13. P. Davis and V. Lin, Public health in Australia and New Zealand. In R. Beaglehole, ed., *Global Public Health: A new era* (Oxford: Oxford University Press, 2003), pp. 191–208.

14. I. Jones and D. Walker, The role of theory in public health. In G. Scally, ed., *Progress in Public Health* (London: Royal Society of Medicine Press, 1997), pp. 57–72.

15. D. L. Weed, Towards a philosophy of public health. *J. Epid. Comm. Health*, **53** (1999), 99–104.

16. A. S. Kessel, Public health ethics education in the United Kingdom: questionnaire survey. *Soc. Sci. Med.* **56** (2003), 1439–45.

17. D. Hunt, Go for ethics – and for growth. *Observer* 19 September (1999), Cash/Special Report: Pensions: 15.

18. R. Cooter, The resistible rise of medical ethics. *Soc. Hist. Med.*, **8**:2 (1995), 257–70.

19. M. Darragh and P. M. McCarrick, Public health ethics: health by the numbers. *Kennedy Inst. Eth. J.*, **8**:3 (1998), 339–58.

20. K. C. Calman and R. S. Downie, Ethical principles and ethical issues in public health. In R. Detels, W. W. Holland, J. McEwen and G. S. Omenn, eds., *Oxford Textbook of Public Health, 3rd edn. Volume 1: The scope of public health* (Oxford: Oxford University Press, 1997), pp. 391–402.

21. J. Fairbanks and W. H. Wiese, *The Public Health Primer* (Thousand Oaks: Sage, 1998), pp. 110–11.

22. S. S. Coughlin and T. L. Beauchamp, eds., *Ethics and Epidemiology* (New York: Oxford University Press, 1996).

23. S. S. Coughlin, *Ethics in Epidemiology and Public Health Practice* (Columbus: Quill, 1997).

24. R. Tong, The ethics of care: a feminist virtue ethics of care for healthcare practitioners. *J. Med. Phil.*, 23:2 (1998), 131–53.

25. G. Weisz, ed., *Social Science Perspectives on Medical Ethics* (Dordrecht: Kluwer Academic Publishers, 1990).

26. B. Hoffmaster, The forms and limits of medical ethics. *Soc. Sci. Med.*, **39**:9 (1994), 1155–64.

27. T. Beauchamp and J. Childress, *Principles of Biomedical Ethics* (New York; Oxford University Press, 1994).

28. R. Gillon, Medical ethics: four principles and attention to scope. *BMJ*, **309** (1994), 184–8.

29. R. Davis, The principlism debate: a critical overview. *J. Med. Phil.*, **20** (1995), 85–105.

30. D. Callahan, Principlism and communitarianism. *J. Med. Ethics*, **29** (2003), 287–91.

31. *Ibid.*, 289.

32. P. Stewart, Academic medicine: a faltering engine. *BMJ*, **324** (2002), 437–8.

33. J. Bell on behalf of the Working Group of the Academy of Medical Sciences, Resuscitating clinical research in the United Kingdom. *BMJ*, **327** (2003), 1041–3.

34. L. Pritchard, Crisis in academic medicine prompts BMA rescue package. *BMJ*, BMA News, 10 April 2004, 1.

Index

Page numbers in *italics* refer to figures. Page numbers in **bold** denote entries in tables and boxes.

Lightning Source UK Ltd.
Milton Keynes UK
08 April 2011

170629UK00004B/2/P